Extension Innovations for Agricultural Development

THE EDITORS

Prof. Baldeo Singh, Ph. D. (Agril.Ext.) and born in 1948, is a Former Joint Director (Extension), IARI, Pusa, New Delhi. He has over 40 years of service for the cause of farmers and rural development as social scientist and contributed significantly for the growth and development of Extension Education discipline. Starting his career as lecturer at C.S Azad University of Agriculture and Technology, Kanpur from 1970 to Jan. 1977, he entered in ICAR service through 1st batch ARS in Feb.1977 and worked in different capacities at Indian Veterinary Research Institute, Izatnagar; Central Institute for Research on Goats, Mathura; ICAR Research Complex for NEH Region, Meghalaya and lastly for over 26 years at IARI, Pusa, New Delhi. He was actively involved in teaching, research and transfer of technology and guided 8 PG and 16 Ph.D. students in relevant areas of research. His keen interest and diligent pursuit as academician fetched him the IARI Best Teacher award 2002. Besides conducting many researches of national importance funded by IARI and ICAR/DST/IRDC, carried out a collaborative action research with Michigan University, USA to train young extension scientists working in ICAR institutes and SAUs as Programme Evaluators. He was also awarded by IARI in 2005 for his outstanding field extension and Extension Management work. He has to his credit more that 300 research and technical papers and about 18 authored/ edited books, 28 training manuals, 1 multi-media DVD and 3 Video films. The Indian Society of Extension Education bestowed upon him its coveted award of 'Dr. K. N. Singh Memorial Award' for outstanding research and 'ISEE Fellow Award' for outstanding contribution in agricultural extension discipline. At present, he is President of Indian Society of Extension Education. He visited three foreign countries *i.e.* Phillpines, Nepal and Pakistan in different programmes.

Dr. Madan Mohan Patel is currently employed as Professor and Head, Department of Extension Education, Rajmata Vijayaraje Scindia Krishi Vishwa Vidyalaya, Gwalior, M.P. He obtained M.Sc. (Ag.) and Ph.D. in Extension Education from JNKVV, Jabalpur M.P. He had 35 years experience of teaching, research and extension. Under his guidance more than twenty five M.Sc. Theses and one Ph.D. thesis have been completed. He has written two books, published more than fifty research papers in repotted journals and edited one book. He has also been honored by Indian Society of Extension Education as "ISEE Fellow Award". Dr. Patel is the life member of seven societies of extension education. He has vast experience in organization of trainings, and other extension activities. He has also organized two national seminars.

Dr. Shobhana Gupta, currently serving as Deputy Director Extension in RVSKVV, Gwalior (MP), did her M.Sc. in Home Science Extension Education and Ph.D. in Agricultural Extension. Recipient of University Gold Medal, she had a continuous brilliant career throughout her education. She has a vast experience of nearly 20 years of working in KVK system. Being a promising, young and dynamic women scientist, she has been awarded on various forums for her enthusiasm towards working for the betterment of rural community. She has to her credit many books, booklets, research papers, popular scientific articles, technical bulletins. radio and TV programme etc. She is life member of many scientific societies and editorial board of various journals of academic interest. She writes from a background of long and varied practical experience as extension worker, teacher and administrator.

Dr. Yagya Dev Mishra is Scientist (Agricultural Extension) in the Directorate of Extension Services, Rajmata Vijayaraje Scindia Krishi Vishwa Vidyalaya, Gwalior (M.P.). Prior to the current assignment, he served as Media Research Consultant at TALEEM Research Foundation (A research arm of Zee Telefilms), Ahmedabad and KVKs under, GBPUA&T, Pantnagar. He has completed Masters' and Ph. D. in Development Communication from GBPUA&T, Pantnagar with specialization in Agricultural Communication. He has more than 13 years of experience in communication research, field extension management and teaching of extension courses at post graduate level. He has published more than 50 research & extension papers in reputed journals & other periodicals and six chapters in various books. He has been bestowed with Young Scientist Award from ISEE, New Delhi and SSDAT, Meerut.

Dr. R.N. Padaria is a Principal Scientist in the Division of Agricultural Extension at ICAR-Indian Agricultural Research Institute (IARI), New Delhi. Extension systems, Climate change adaptation, Farmer participatory research, Indigenous knowledge system, Socio-economic and environmental implication based research (Big dams and transgenic agriculture), and gender empowerment have been his major areas of research. He has more than 60 national and international publications. He has made significant contributions in postgraduate teaching and research programme at ICAR-IARI and received several professional awards.

Extension Innovations for Agricultural Development

— Editors —

Baldeo Singh

Madan Mohan Patel

Shobhana Gupta

Yagya Dev Mishra

R.N. Padaria

BIOTECH

2021

BIOTECH BOOKS®

ISBN 978-81-7622-395-9

Published by: **BIOTECH BOOKS®**
4762-63/23, Ansari Road, Darya Ganj,
New Delhi - 110 002
Phone: +91-011-23262132
E-mail: biotechbooks@yahoo.co.in

Typeset at:**Classic Computer Services**
Delhi - 110 035

Digitally Printed at: **Replika Press Pvt. Ltd**

PRINTED IN INDIA

राजमाता विजयाराजे सिंधिया कृषि विश्वविद्यालय
रेस कोर्स रोड, ग्वालियर (म.प्र.) – 474002
Rajmata Vijayaraje Scindia Krishi Vishwa Vidyalaya,
Race Course Road, Gwalior (M.P.) - 474 002
(An ISO Certified 9001:2008)
Tel: 0751 – 2467673, Fax: 0751 – 2464141, E-mail: vcrvsaugwa@mp.gov.in

प्रो. अनिल कुमार सिंह
कुलपति
Prof. A.K. Singh
Vice-Chancellor

No./VC/2016-17/ 1695
Date: 28/08/2016

Foreword

Agriculture sector has registered an appreciable growth in the country with adoption of innovative technologies. Augmenting productivity, production and enhancing farmers' income are in the main agenda for all stakeholders associated with agricultural activities in spite of the challenges posed by globalization and climate change. Concerted efforts are being made to involve farmers as effective partners in technology generation and utilization. Despite the remarkable growth in service sector during past two decades, the agriculture sector is still considered and will remain the backbone of Indian economy. It is, therefore, high time that the National Agriculture Research System in the country resets its priorities to work and support local initiatives for sustainable food and nutritional security.

It is a fact that farmers are the sole decision makers for adoption of new technologies and therefore, their participation has proved as an effective tool in development of need based technologies for sustained growth in agriculture.

I am happy that by recognizing the importance of innovative extension methodologies, this team of scientists has brought out this edited book entitled "Extension Innovations for Agricultural Development". I am confident that this publication will be of immense value to scientists, farmers, extension personnel, administrators, policy makers and all the stakeholders, who are associated with agricultural development sector.

It also gives me a great sense of satisfaction to see that the publication is an output of the deliberations that took place during the three days National Seminar of Indian Society of Extension Education held at RVSKVV, Gwalior (M.P.) from 26th to 28th February, 2015. It has effectively captured the experiences related to extension innovations and methodologies for market led agricultural growth and development.

I congratulate all the contributors in bringing out this topical publication.

A.K. Singh

Preface

The country has long back moved from subsistence agriculture to intensive agriculture with higher production and profitability. Recently Government of India shifted its focus and emphasis to commercial agriculture and agricultural exports. The science and technology development in agriculture has opened fabulous opportunities for income and employment generation. This demands a market-led agriculture strategy to facilitate development of local capacity to exploit the opportunities of scientific agriculture for better socio-economic order. There are problems of surplus as well as sustainability. Therefore, there is need to promote diversification, value addition & processing and innovative marketing of agricultural products nationally and globally. The farm households, environment, satisfaction and prosperity are the top most agenda for all round development of the rural community. The empowerment and development of rural people and sensitization of research and extension systems about their roles and contributions are important for sustainable growth in agriculture. The present agricultural production system is highly competitive, particularly when a number of events are taking place simultaneously including issues of marketing. Agribusiness management including efficient marketing systems and risk management are the keywords of today's agriculture.

Agricultural Extension, in the present scenario of rapidly shifting world has been recognized as an essential device for transferring information as an input into contemporary farming. The selection of an extension method depends on number, educational status and location of the target audience and the time available for communication. Each of the methods has both advantages and limitations. The extension professional has to choose a particular method or combination of methods according to the needs of the situation. With the advent of new technologies, agriculture has gained momentum and is poised towards horizon of new possibilities like e-connectivity, cyber extension etc. Innovative approaches have catalyzed the penetration of knowledge based awareness to the grass-root level bringing out radical transformation in thinking and approach for development of agriculture in general.

In the current assemblage, efforts have been made to present authors' contributions in their original form with some necessary editorial amendments required for publication in this manuscript.

The editors convey their sincere thanks to the authors for their contributions and optimistic response in submission of their documents.

Sincere gratitude from editors to the ICAR, New Delhi for providing financial aid for organizing the ISEE - National Seminar and as an outcome the pooled experiences and ideas of the professionals are available in this document for ready reference to the planners, extension professionals, service providers and other stakeholders in agricultural and rural development.

The editors express special gratitude to Prof. A. K. Singh, Hon'ble Vice Chancellor, RVSKVV, Gwalior for his kind support and encouragement.

Baldeo Singh

Madan Mohan Patel

Shobhana Gupta

Yagya Dev Mishra

R.N. Padaria

Contents

Chapter 1

Extension Innovations and Methodologies for Market-Led Agricultural Growth and Development

Prof. Baldeo Singh

Former Joint Director Extension (IARI) and President, Indian Society of Extension Education, IARI, New Delhi

Agriculture has been the anchor for achieving the socio-economic growth. While food security still remains a matter of concern, the present agricultural scenario has resulted into new priorities and calls for meaningful role of extension education to exploit the opportunities for the desired gains. The science and technology development in agriculture has opened tremendous opportunities for income and employment generation. Today, a new paradigm of agricultural development is fast emerging and demand for agricultural technology are changing and diversifying. The overall development of rural areas is expanding in new directions, old ways of delivering important services to the people are being challenged and traditional societies are being transformed into knowledge societies. Markets are now the driving force for agricultural and rural growth, depending on effective information support *i.e.* what people want, at what price, where to get it, and who can supply it. Developing the capacity to generate, absorb, disseminate and protect knowledge and exploit it as a powerful tool is necessary to derive societal transformation.

The ISEE national seminar on 'Extension innovations and methodologies for market-led agricultural growth and development' is organized to platform academicians, researchers, thinkers, practitioners, policymakers and innovative farmers for deliberating all such issues threadbare under the following themes:

1. Linkages and institutional innovations for market led agriculture
2. Extension innovations, methodologies and experiences in technology transfer, quality up-gradation, processing, value addition and marketing
3. Rural entrepreneurships for market driven agriculture and livelihood systems
4. Policy initiatives for market led rural economy
5. ICT for agri-business management and marketing
6. Gender mainstreaming - extension imperatives for market led agriculture
7. Harnessing of ITKs and farmers' innovations for agro-marketing systems

Agenda for Agricultural Extension

The agenda of extension initially was to increase food production. Even-though improving productivity to meet the demands of food and nutritional requirements of growing population will continue to remain as one of the main agenda of extension, the concerns for receiving the greater attention of extension in sustainable production are environmental protection, reduction of poverty, conservation of natural resources, improving the profitability of farming through agro-processing and value addition, and marketing, improving the participation of stakeholders in decision making process and improving the efficiency of extension organizations through adoption of improved management practices.

Indian agriculture is getting more and more market and export oriented. With rapid changes in agriculture, we need to maintain pace and prepare to meet the emerging demands through our meaningful research, training/education and transfer of technology. There is a need to identify the competitive advantages that our research and extension systems have and exploit their potentials in order to retune our programme and strategy to respond to the present situation. The following considerations are important:

1. Maintaining and consolidating the gains made in agriculture and continuing to enlarge levels of potential productivity and profitability with eco-friendly technologies
2. Extending agricultural gains in integrated manner on integrated farming systems development to the areas so far largely deprived, especially in rain-fed and dry areas, where the farm and non-farm technologies have made only a limited impact.
3. Increasing emphasis on post harvest technologies, value addition and allied technologies on commodity and area basis for entrepreneurship development to take the fullest share of exports.

Emphasis on Integrated Farming System Approach

Basically, the agricultural development is a holistic concept encompassing both natural and human resource development. The National Policy on Agriculture envisages promotion of sustainable agriculture through a regionally differentiated farming systems approach, development and transfer of technology, improvement of input use efficiency, incentives for agriculture, strengthening of infrastructure, risk management and management reforms.

The sustainable development has acquired a sense of urgency and demands for strategic action to address the scenario in an integrated manner. The integrated farming system approach is helpful in utilizing local resources, blending modern technologies with indigenous practices, enhancing production and employment potentials including entrepreneurship development. The farming system is a unique and reasonably stable arrangement of family enterprises that the household manages according to its physical, biological, economic and socio-cultural environment in accordance with household's goal, preferences and resources.

In view of above and to promote market-led agriculture, the following extension strategy is considered appropriate:

1. Strengthening National Extension System with Robust Structure and Organization

Usually extension has been used to meet expanding demands for food and to cope with declining availability of various resources. Its educational role in terms of problem diagnosis, effective transfer of technological messages though educational aids, feedback of farmers' response to technology and development of effective linkages with different agencies to solve problems of farmers is almost missing. We need a strong and responsive extension system, which can cope up the challenges and meet the needs and aspirations of rural poor. The very design and approach of extension is required to be resurrected and strengthened in lines of **community development approach** to attend multifarious demands of the farming community. It is essential to infuse professionalism, responsiveness, accountability, missionary zeal and sound managerial abilities among extension functionary to handle the farming constraints and challenges.

There is a need to revamp public extension network with leading role and involve selectively the private sector particularly in the post harvest processing, value addition and marketing. The extension system should be reoriented with respect to required human power (gender-wise) and adequate logistic support of modern information and communication technologies, adequate budget, transport facilities etc. to cater emerging farming needs. The entire functionaries are required to be trained in newer and relevant areas of behavioral dynamics and methodologies, focusing upon market extension to help farmers with market information as well as upgrade their marketing skill.

2. Revitalizing Extension System of SAUs and ICAR

The extension units of SAUs and ICAR institutes must be manned by extension specialists with adequate human power (male and female) strengthening, budget provision, vehicles for mobility, decentralized management and creation of single window delivery outlets, audio-visual aids, IT facilities, mobile cum- exhibition and video vans, etc. **Training** should be emphasized as important function of SAUs/ICAR institutions besides research, education and extension. Hence, Advanced Training Centers (ATCs) may be created in selected SAUs/ICAR Institutes. The ICAR led previous front-line extension projects like Lab to Land Programmes (LLP) and Operational Research Project (ORP) should be reintroduced for socio-economic betterment of small and marginal farmers. The KVKs should be equipped and should work as role model and light-house for agricultural development with new technologies and need and skill based trainings for promoting entrepreneurships, rather than just a center for technology testing.

The facts about the structure and professional set up of extension system in SAU/ICAR including KVKs reveals that the people not having professional extension degree manage most of the extension affairs. It is essential to introspect and deploy extension specialists for lead role through policy design and intervention.

3. Public-Private Partnership

The integration of research, extension and client systems for sustainable development have assumed greater importance today. The private extension works on the specific areas and can meet the demand for the specific information more efficiently than public extension system because of its network. Therefore, the private services need to be properly channeled and utilized to support the main CD extension system, particularly for post harvest management, entrepreneurship development through small-scale cottage infrastructure and export promotion. In this context, the public-private partnership has been expressed explicitly as follows:

- ✰ Promotion of Agribusiness through development of Agribusiness Center
- ✰ Establishment of Agro-Clinics

4. ICTs Infusion (Info-tech and Agro-biotech)

ICT diffusion and infusion in agriculture provide necessary digital opportunities or advantages for productivity increase, income generation, decrease in national disparity and improving linkages with the market. Net working of people and networking of information are required for reaching the unreached through public extension system. Preparation for future development and communication among masses require an extensive thinking regarding information technologies and convergence with mass media including the Internet. An important development is the emergence of e-Commerce. More and more companies are setting virtual showrooms, which can be assessed by any users.

5. Intensification of Field Extension Programme

Besides conventional methods of demonstrations, exhibition, excursions, growers-scientists interface, field-visits, commodity shows and competitions at block, district, state and national levels and innovative approaches of FARMERS' FIELD SCHOOL, FARMER-to-FARMER communication and rewarding the pioneer adopters in public gathering may be effective in dissemination of knowledge and infusion of competitive spirit among the growers.

6. Group Approach Extension

The viability of small and marginal production system needs to be invigorated with group action. Emphasis should be laid upon promotion of self-help groups; commodity groups; farmers' associations, forums/clubs etc. to enhance their operational and bargaining capabilities.

7. Agricultural Marketing and Trading

a. Organized retail involving farmers/their associations can help reduce wastage that currently characterizes farm production. The studies indicate that about 50 per cent of the population, below 30yrs of age, is aspiring for higher production, consumption and innovation. Involvement of these groups in retailing of food and agricultural products by extension design to build their capability and financial investment/infrastructure support may help bring about a transformation in farm sector to improve production, productivity and quality. The challenges arising out of fragmented holding, limited credit flows and uncertain market conditions may be tackled by such extension design in integrated manner.

b. **A systematic area planning** in concentric mode with due emphasis on important commodities in production – food value chain is an appropriate strategy. The comprehensive area planning is required to develop agriculture with vertical and horizontal links of concentric zones so as to develop block as hub or nucleus commercial zone primarily of tertiary sector and centrifugal belts as feeding zones of secondary and primary sectors. Linking small farmers to processing industries and fast food chains would be a good proposition. The food production and retail game has the potential to turn into a win-win situation for all stakeholders-growers, aggregators, retailers and consumers – if it is played with social responsibility. The priority need is to thoroughly study the existing retailing of agricultural products and promote it on three tiers basis. **First tier** - high-tech diversified agriculture, processing, value addition and trading at District level; **Second tier** - integrated agriculture, processing, value addition and retailing at Block level and **Third tier** - commodity based integrated agriculture, pre-processing and petty retailing at Village *Panchayat* level. While retailing at first tier should be promoted in partnership of corporate sector, but the retailing at second and third tiers restricted to rural entrepreneurs to create employment market and utilize local market avenues and also to promote integrated farming system for sustainable production and higher profitability. The financial support at these two lower levels is essential.

c. **The key areas** of food retailing are required to be looked into. These are storage, transportation, segmentation based on class of buyers, packaging etc. in retailing. The partnership issues need critical examination to delineate roles and responsibilities of different players while safeguarding the local interests. Yield, price and market risks remain major constraining factors for smallholders' participation in retail marketing. Smallholders are risk averse and their effective participation in regional, national and global markets requires a mechanism that could absorb such risks. There is a need to promote entrepreneurship with financial and infrastructure support by corporate on benefit sharing basis.

d. **The contractual farming** being promoted without clear arrangements and institutional and legal mechanisms to resolve disputes can exploit smallholders and private sector may lose money if farmers do not meet their supply obligations. Forming producer cooperatives is the way to make small farmers credit worthy for organized farmer-run supply chains. Smallholders have shown that they could meet the quality requirements of markets if adequate support in terms of infrastructure and training is given to monitor the quality standards.

8. Participatory Development with local Leadership

a. An integrated farming systems approach for participatory development on watershed basis has become essential. Wherever the people did the problem articulation, programme development, implementation, monitoring and evaluation, the programme had a long lasting positive impact. The need, therefore, is to have a correct understanding of the people, their social systems, gender needs and interests and local environment; and plan, execute and evaluate their development programmes through local leadership. Local leaders can mobilize the community better and make them participate in the development programmes. A strategy must be devised which should purposefully guide the promotion of participation within the project emphasizing the development of local leaderships

b. We have to look for effective methodology to transfer the technologies at a much faster rate so that the varieties or breeds, products, techniques developed with the required traits can be transformed to the field without much loss of precious time. In this context, the IARI initiative during the year 2007 to develop **'Village based integrated models for market-led agriculture'** in different parts of the country in collaboration with SAUs/ ICAR institutes is noteworthy. In this effort, progressive farmers were linked and utilized for farmer to farmer technology transfer, which helped in wider spread of technology adoption.

We need to promote a mission mode project on integrated farming system approach so that farmer's income can be enhanced by promoting agro-based enterprises such as bee keeping, rearing animals, mushroom cultivation along with crop cultivation. The major demand today is of energy, which can be met by developing an **energy grid/Energy Bank** harnessing renewable energy/biogas from different sources. The villagers should be encouraged to develop **Water bank** by harvesting whole of the rain water and then judiciously using it on benefit sharing. Each Village Pachayat should have a **Seed bank** by producing enough good quality seeds and planting materials according to their need and requirements. **Manure bank-** the cattle dung and other bio-mass are precious village resources add plant nutrients and improve soil health and physical properties. The scientific management of biomass can also provide renewable energy and reduce pollution. This way if we have to revive agriculture and we can develop resource base at village level, while enhancing production and productivity.

9. Gender Empowerment and Mainstreaming

Paradigm for practicing agriculture having farmwomen in focus has to be restructured according to resource potential and market feasibility. The challenge for participatory development lies, primarily, in making the women a part of the mainstream development process. Studies conducted over three decades have established the significant role of women in varied agricultural tasks. Having vast knowledge of the genetic resources, they play a crucial role as custodians of the genetic diversity and the related knowledge on the varieties and their uses for food, medicine etc.

References

Gautam, R. C. and Sharma, A. R. (2003). Alleviation of hunger and poverty through integrated farming systems in fragile eco-system. Indian Farming, 53(7): 35-39

Prasad, C., B.N. Chaudnary and K.S. Krishnan (1990). Impact of Agricultural Modernization on farm women in rainfed areas: an Indian case study in "Extension Designs and Gender Issues in Rainfed Agriculture" published jointly by Indian Society of Extension Education, IARI, New Delhi, India and International Fund for Agriculture Development (IFD), Rome Italy.

Rao, C.H. Hanumantha (2001). WTO and Viability of Indian Agriculture, economic and political weekly, September 8th.

Ruthenberg Hans (1971). Farming systems in tropics. Clarendon Press, Oxford.

Shaner, W. W., Philip, P. F. and Schmethi, W. R. (1982). Farming systems research and development: Guidelines for developing countries. Boulder Co.: Westview Press.

Singh B. and Poonam Sharma (2001). Integration of Rural Women in Development through empowerment", Indian Research Journal of Extension Education, Vol.1, No. 2, July 2001.Schultz, T.W. (1990), 'Restoring Economic Equilibrium: Human Capital in the Modernising Economy". Oxford, Basil Blackwell.

Chapter 2

Futuristic Agricultural Extension Models and Approaches

K.L. Dangi

Department of Extension Education, Rajasthan College of Agriculture, Maharana Pratap University of Agriculture and Technology, Udaipur, Rajasthan

Extension, as the organized exchange of information and the purposive transfer of skills, is a rather recent phenomenon. Obviously, transfer of information and skills has existed since the emergence of permanent agriculture. Today's practice is different in that the process is dominated by organizations, and its scope has extended from disconnected local events to a complicated, large-scale, and even worldwide activity. Main actors within the extension system are the members of rural communities, extension and other development personnel, researchers, and staff of commercial or public service and support organizations. Empirical evidence shows a variety of forms in which interaction among these groups is institutionalized.

The variety of forms suggests a similar variety of goals, and either could be used to classify extension approaches. In practice, however, one finds an almost inseparable mixture of goals inhibiting a clear-cut classification. It seems more appropriate, then, to use a broader category, namely, selectivity with regard to clientele, and treat the respective goals as a continuum. The two end points of this continuum would be marked as technology transfer and human resource development, suggesting either a rather narrow technical or a broader socioeconomic view of development.

Until the end of the eighteenth century, farming techniques developed gradually and steadily over centuries with few qualitative leaps. Colonialism and imperial expansion introduced innovations - the spread of maize, tobacco, potatoes are striking examples - but experimentation and dissemination of knowledge were basically at the local farm level. The rise of agricultural sciences has induced dramatic changes in this respect. Increasingly, new technology has been created outside the actual farming sector by public sector research organizations. More recently, private firms in industrialized economies find agricultural technology research and development a highly profitable business.

For decades the research-extension-farmer linkage, especially in developing countries, was based on a rather simple model. In order to achieve development, "modem" research results had to be transferred to the "traditional" farmer, and extension seemed to be the appropriate means to do so. The general faith in science and the commitment to modernization led to discrediting indigenous knowledge. Although this view is still held by many administrators, researchers, and extension agents, it is now being seriously questioned.

Farming systems research and the "rediscovery" of farmers' knowledge (Chambers, Pacey, and Thrupp, 1989) have shown that "improved technology is a package of inputs and practices that usually comes from many sources" (UNDP, 1991, p. 2). The reexamination of the conventional view on agricultural knowledge cannot, however, result in questioning the important role of research as the source of new technology. For developing countries, one observes that the accelerated growth and spread of problems - such as the degradation of marginal land - surpass the problem-solving capacities of the local population. What is called for is a setting of new priorities and the building of knowledge systems based on problem solving rather than on information transfer.

The nodal points of Indian prosperity are:

i) Peace, security and national unity

ii) Food and nutritional security

iii) Jobs for all

iv) Knowledge

v) Health

vi) Technological and infrastructure

vii) Globalization

viii) Good Governance

ix) Work values.

Experiences from early attempts and present situation of agriculture clearly show that no single model or mode will give good results in all situations. Private sector involvement hass to encouraged in extension and services support such as agri-business centres, agri-clinics, linkage between KVKs of ICAR and State/ District extension services together with that of private sector/NGOs involved in Agricultureal Extension. It is therefore, imperative that we have to look for different alternatives or models depending upon the clents, objectives etc, and the dichotomy include:

i) Public vs Provate

ii) Government vs Non- Government (Voluntary)

iii) To-down(bureaucratic) vs Bottom- Up/Horizontal (participatory)
iv) Profit vs Non- Profit/service
v) General vs Sectoral
vi) Multipurpose vs single/Special
vii) Technology – Driven vs Need/demand – Driven etc
viii) Free versus cost-recovery
ix) Commodity Based Extension.
x) Extension as a Commercial Service
xi) Client-Based and Client-Controlled Extension.
xii) Farmer First
xiii) Panihari- Extension
xiv) Gwala- centred Extension
xv) ICT mediated Extension.

I am of the firm opinion that, if these above mentioned modes/models of future agricultural extension are made use of, we can expect spectacular sustainable development in agriculture.

I suggest following specific Futuristic Agricultural Extension Models and Approaches.

1. Commoditization of Quality Extension System

Introduction

Agriculture has been the anchor for achieving the socio-economic growth. It accounts for about 1/3rd of annual Gross Domestic Product (GDP) of the national employment. During last five decades, the country has witnessed spectacular progress in agricultural production increasing its total food production from 51 million tones in 1950 to about 206 million tones in 2005. it is testimony to resilience of Indian Agriculture and manifestation of hard work of farmers, extension workers, scientists and administrators. The strong agricultural research and education system coupled with appropriate extension support have helped in achieving the food sufficiency to meet the increasing food demands despite growing population. The challenges posed by agricultural globalization, revolution in information and communication technology, rising population and consequent demand for sustainability have prompted to have a re-look on the existing extension system with proper direction.

A good number of approaches are available for agricultural and rural development. With the help of existing statistics we can say that task ahead of extension system is gargantuan (gigantic). The estimated population of the country is 1009 millions, net cultivated area is 141 mha, gross sown area is 203 mha, gross irrigated area is 114 mha and gross irrigated area as percent of gross sown area is 56 percent (Alagh 1996). Coupled with managing, maintaining and sustaining seven pillars of integrated intensive farming system; soil health care, water harvesting and management, crop and pest management, energy management, post harvest management, choice of crops and other components of farming system, information skill, organization and management empowerment need special attention by extension system (Swaminathan 1996).

Extension Services through Extension System

Since long experiments are going on with the extension system in India. Extension system at the national level was first introduced in the year 1952, but before that extension efforts were going on for the upliftment of rural life. A brief list of extension efforts will make a bench mark for development of extension system in India. There were two distinct types of efforts made to do the welfare of the rural people. First some benevolent persons and private agencies made some efforts to improve rural life, second government efforts were made to initiate some projects to solve the emerging problems. List of extension efforts is presented below with the year, name of initiator and title/place of experiment.

A bird's eye view of the various projects/programmes taken up throughout the nation may be with internal resources or with the help of external resources. The outcome of the programme/projects can be summarized in the perspective of relationship of the project/programme with the farmers. According to Rivera (1989) there were four types of relationship that existed with the farmers. The relationships that existed were,

1. Take it or leave it
2. Take it or demand different packages or programmes
3. Take it .... Or else
4. Take it or run away

The various systems approaches, the type of system and the relationship to farmers is presented in tabular form.

Extension approaches and systems and their relationship to farmers (Rivera 1987).

Sl.No.	*System Approaches*	*Type of System*	*Relationship to Farmers*
I	Top down delivery services	Conventional, T and V, University organizations, technical innovation, integrated agricultural developmental programmes	Take-it-or-leave-it
II	Participatory acquisition system	Farm information dissemination, Farming System Research and Development	Take it or demand different package or programmes
III	Contract farming system	Commodity development and production, Commodity focused	Take it.... Or else
IV	Rural development extension approaches	Community development cum extension, Integrated rural development programmes	Take it or run away

National arrangements and systems approaches for extension and their relationship with farmers. (Rivera 1987).

Sl.No.	National Arrangement	System Approaches	Relationship to Farmers
I	Public Ministry of Agriculture	Delivery service (non compulsory)	Take-it-or-leave-it
II	Dual Ministry of agriculture and farmers associations (dual control over extension service)	Participatory (Shared responsibility)	Take it or demand different package or programmes
III	Coordinated Ministry of Agriculture (Parastatal) may be public or private		
IV	Private For profit: Domestic enterprise cooperatives multinational enterprises Membership: farm associations Non profit: NGOs		Take it.... Or else

Farmers' Perception of Different Approaches

The earlier extension efforts taught the field extension functionaries as well as extension educators how the farmers perceive the extension initiatives. All efforts made so farm have been emphasizing on participation, may be directly on indirectly. But all these efforts put together could achieve little as far as 'participation' of community is concerned. The study conducted in this regard reveals the farmers perception of different approaches (Hagmann *et al.*, 1996) as under.

- ☆ Forceful methods were followed
- ☆ Only a few people could benefit
- ☆ We were told to do things without questioning
- ☆ There was no dialogue between farmers and extensionists
- ☆ There was little cooperation among farmers
- ☆ Extension agents treated our fields as if they owned them

Systems Approach of Extension

Extension systems have been explained by different authors (Oxenham and Chambers 1978, Orivel 1981, Pickering 1987, Ray 1985, Widemann 1987), but there are contradictions among typologies and some confusion of terms. Pickering (1987) distinguished the following systems.

- ☆ The commodity focused approach in extension designed to facilitate production of a single crop
- ☆ The community development cum extension approach incorporating a broad definition of functions of the extension agent, which tends to dilute the agents specific agricultural extension responsibility

- The technical innovation centered approach set up to transfer technology from outside to the farm sometime specifically to sell a number of technical innovations
- The Training and Visit system approach, organized to serve farmers by mobilizing the extension system, as well as its linkage with research system, through regular visits by agents to farmers any by regularizing of agent training.

Lele (1975) explained extension broadly under two major directions *viz.* (a) the take it or leave it approach where farmers are free to accept or reject development innovations and (b) the contract farming approach. In the later farmers are granted a license to produce certain commodities on the condition that they use a particular innovation and follow the project guideless set down by the extension organization.

Oxenham and Chambers (1978) and Orivel (1981) described another approach as representative participation, wherein both local government and farmer associations were involved in controlling the system. Axinn (1987) classified approaches by a point of control: the delivery approach (top down supervisory and supply driven) and the acquisition approach (bottom up, farmer determined, participatory and demand driven). Ray (1985) classified three approaches. The directive (top down delivery systems), the participatory (systems involving farmers participation) and the contractual model (systems where farmers contact directly with public agencies or private companies to receive extension service).

Need for System Revival through Comunitization

In a sampling survey conducted in 1977 in Latin America it was observed that 82 percent of the researchers felt that extension personnel made too little effort to learn and to transfer the technologies generated and 72 percent of the extensionists felt that the new technology was not acceptable to small holders.

The World Bank (1985) identified the gaps between research and extension systems as the most serious institutional problem to be overcome in developing an effective research and extension system. Uphoff and Whyte (1983), Venkatesan (1985), Johnson and Clear (1986) and the Asian region Workshop of the Training and Visit System of Extension (1982) also noted the need for strong Research and Extension linkage. The linkages as a weakest part of R and E system has also been emphasized in the literature and the reason for it are assigned (Cumings, 1981, Fernandez, 1981; Mosher, 1978; Maalouf, 1983).

ICAR conducted a study revealed that overall adoption of the agricultural technologies released by research institutions since independence. The results of the study are alarming. The study revealed that overall adoption of the agricultural technologies released by research institutions since independence was only 30 per cent. The most common reasons quoted were ineffective extension, inefficient supply of inputs, credit support, lack of market infrastructure, farmers' lethargy and indifference for the knowledge practice gap. But in reality the research says the non adoption was due to technologies are not economically viable, not operationally feasible, not stable, do not match farmers needs and not compatible with overall farming system (Das 1996).

With the experience of various projects/programmes that were implemented since 1901 till date it is evident that the projects/programmes achieved remarkable results to make rural life better. The increase in production and productivity are significant, but still there is lot of scope left to be harnessed. In the wake of increasing awareness among consumers regarding the quality of the agricultural produce, food security and nutritional security, quality of production process including research and extension systems are to be reoriented. The point of intervention at this juncture shall be communitising the extension services. The concept of communitization was mooted by Sri. R. S. Pandey, former chief secretary of Nagaland.

The reform imperative concept 'comunitisation' took birth in due course of time. The community (people) developed a typical attitude that everything was owned by government. There was little sense of belonging, ownership of responsibility to public institutions and services. The most common behaviour of common people like disregarding, mutilating or even destroying public properties was increasing at national level. With the mode of delivery of goods and services to the communities through various projects/programmes enlisted above. Even today people expect government should deliver quality services in everything. Against this background it can be concluded that there is deep resentment and frustration among people at perceived failure of government in service sector. It is in this context communitisation of extension system becomes a relevant concept to take care of the issues pertaining to attitude of community.

Communitisation connotes partnership between the government and community involving the transfer of ownership of public resources and assets, control over service delivery, empowerment, decentralization and delegation and building capacity. It is based on triple "T" approach, trust the user community, Train them to discharge their new found responsibilities and transfer governmental powers and resources in respect of management.

The inevitability of communitisation can be explained due to the factors listed below:

- ✰ Community life reflects rich social capital
- ✰ There exists a strong community bond in traditional institutions which are organized, accountable, effective and participatory.
- ✰ Systems perform better when traditional knowledge and wisdom are involved in functioning.

The public extension system needs change for better services by structural reforms and improving function through transferring the ownership to and sharing responsibility of its management with community and by building synergic relationship between community and the public extension system. This can be achieved through communitisation of public extension system.

The Modus Operandi of Communitisation

As a first step to initiate communitisation process; it should be conceptualized and a series of consultation needed at senior state government officials. The process to become operational it needs the approval from the peoples representatives *i.e.* approval by the state cabinet. Then the cabinet has to decide the areas in which the process of communitisation is essential and it should be taken on priority basis. The areas like education, agriculture, health etc. can be taken up on initial priority basis. The government should turn its face towards village. Formation of village level Committees is the first step of communitisation. Then comes delegation of powers and responsibilities to those formally designed Committee. These comities will be having the powers of salary disbursal and committee is empowered to say 'No work no pay', it will be having the control over the attendance of staff and grant of leave. The committee also looks after maintenance of buildings and assets, purchase of essentials etc.

The communitisation shall be effectively conceptualized and implemented only when the community for which it is being planned and designed approved it. The approval of community should be in the form of participation of community in its real sense. It does not mean that the earlier efforts don't have participation of community. There was participation but with the difference. The community didn't

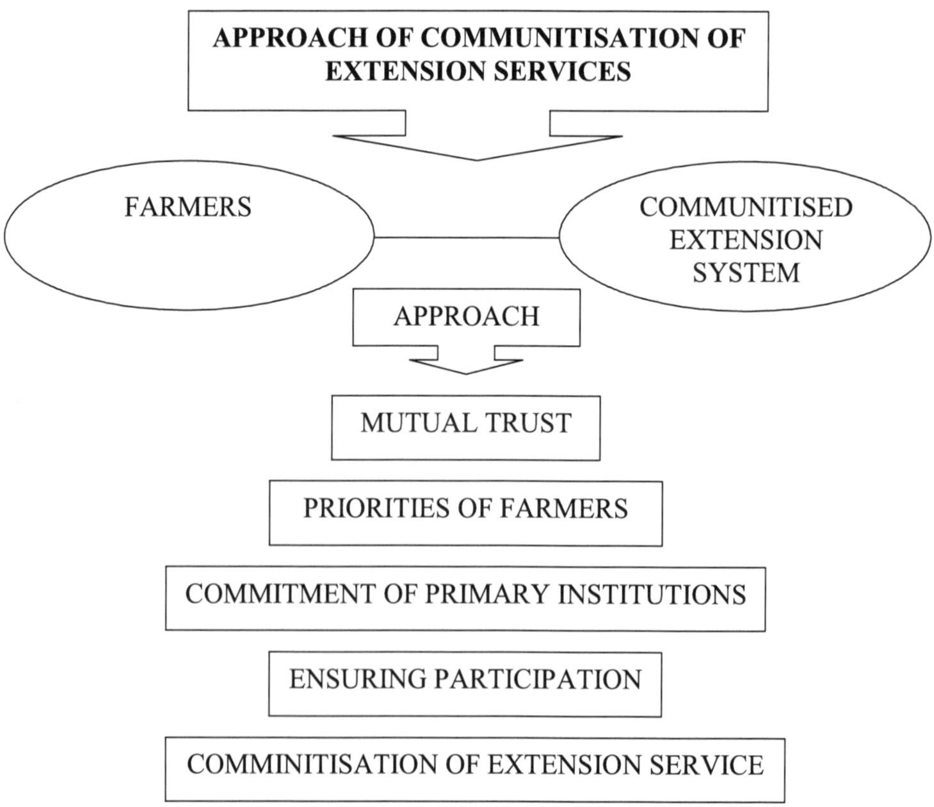

have any decentralized power to execute. Enough powers shall be provided in the concept of communitisation to communities to deliver desired results. This is the striking difference between the proposed and already experimented approaches of extension.

The mutual trust between the farmers and extension worker must be established and priority demands of farmers should be made at the beginning of communitisation process which will satisfy the wider objectives of communitisation through participation.

Communitisation through Participation

Diffusion was the dominant framework in the modernization era (end of World War II to early 1970s) while participation only played a supplementary role in development. In the post modernization era (late 1970's to early 1990's) diffusion was still prevailing but it also faced challenges from the gradually popular participatory approach, which was mainly applied at grassroots level. In the current era of discourse (late 1990's till date) the idea of participation is widespread in the field (Chang 2006).

The following situations force us to think twice before choosing any alternative. The statistics which give the picture of about 70 percent of farmers as small and marginal and about 20 percent of farmers have access to about 80 percent of land and other resources, compels us to communities the extension services which can take care of majority who are in need of critical extension inputs. The farming community which is real minority (big farmers) can have access to extension services by various other means. The majority community in farming (small and marginal farmers) is in need of communitsed extension services. This can only be achieved through participation of beneficiaries. Against this backdrop the concept of participation is discussed.

What is Participation? The essence of the participatory approach to community development is embodied in a poem by Chinese philosopher Lau Tse. "Go and meet your people, live and stay with them, work with them. Begin with what they have, plan and develop from what they know, and in the end, when the work is over, they will say: "**We did it ourselves**"

Participation is a concept with multiple faces. It can refer to stakeholders involvement in the planning or implementing phase, the open and continuing dialogue between program practioners and receivers, or the collective decision making process (Cohen and Uphoff, 1980; Fuglesang and Chandler, 1986; Nair and White, 1994). These various definitions may emphasize participatory activities in different dimensions. Nevertheless, most of them share three fundamental elements: empowerment, dialogue and collective action (Chang 2006).

WHO (1982) defined participation as the process by which individuals, families or communities assume responsibilities for their own and communities development. Cohen and Uphoff (1977) Participation includes peoples involvement in decision making process in implementing programmes and their sharing in the benefits of development programmes and their involvement in the efforts to evaluate such programme.

The rationale behind participatory process of implementing community-level extension services are that, at the end of the service the community can continue alone and it can be called as the Communitised Empowerment Approach.

According to Biggs (1989) farmers' participation can be sought in four different ways contract, consultative, collaborative and collegial.

Participation means more than consultation. It includes providing government information to people in ways that make it accessible and relevant and creating opportunities for people to be actively involved in the design and delivery of government policies and services.

According to Hagmann *et al.*, 1996 Characteristics of Participatory Approach are

- ☆ Everyone can benefit as all are free to attend meetings
- ☆ There is a dialogue
- ☆ The process is well explained
- ☆ Farmers are drivers
- ☆ Farmers are being treated as partners and equals
- ☆ No discrimination against poor/rich, educated/uneducated
- ☆ Farmers are given a choice of options
- ☆ Extension agents pay attention to farmers and take time to find solution to farmers

The Most Important Aspect of Participatory Approach

- ☆ It helps to work cooperatively
- ☆ Farmers practice with enough knowledge of why they should do so
- ☆ Learning from others through exchange visits/learning through sharing
- ☆ It can help farmers to develop the ability to encourage each other in farm activities
- ☆ Encouragement to practice through various options
- ☆ It mobilizes large number of people
- ☆ The approach brings about desirable techniques through participation
- ☆ Farmers are free to ask for advice
- ☆ The dedication of modern extension agents/researchers has psychological impact
- ☆ It will bring about development in the area
- ☆ It is very effective in sustainable development.

Participation Covers the following Three Levels

Information Delivery

information that is relevant, of sufficient quantity and quality- such as timeliness, accuracy, from authenticated sources, etc. – to enable people to make informed decisions when transacting with government.

Consultation

A two-way relationship with feedback sought from people on issues defined by government. Such views are sought to inform decision- making.

Active Participation

Active engagement or relationship with the public and businesses with opportunities to define the process and count of policy-making and service design/ delivery.

A system is usually through of as a group of constituents plus the relationship between them plus their attributes (Casata and Asante 1979). Against this backdrop the participation of community in extension system becomes pivotal. The question to be answered is how to ensure participations? Should it be mandatory, voluntary or optional? Which mode shall would better results is the issue to be addressed. The following principles of participation (www.worldbank.org) help us to find the answers to the questions asked.

Principles of Participation

- ☆ **Inclusion** – of all people, or representatives of all groups who will be affected by the results of a decision or a process – for example a development project.
- ☆ **Equal Partnership** – recognizing that every person has skill, ability and initiative and has equal right to participate in the process regardless of their status.
- ☆ **Transparency** – all participants must help to create a climate conductive to open communication and building dialogue.
- ☆ **Sharing Power** – authority and power must be balanced evenly between all stakeholders to avoid the domination of one party.
- ☆ **Sharing responsibility** – similarly, all stakeholders have equal responsibility for decisions that are made, and each should have clear responsibilities within each process.
- ☆ **Empowerment** – participants with special skills should be encouraged to take responsibility for tasks within their specialty, but should also encourage others to also be involved to promote mutual learning and empowerment.
- ☆ **Cooperation** – co-operation is very important; sharing everybody's strength reduces everybody's weaknesses. The community is responsible for implementing the program activities as specified by the government; Community members attend project planning, implementation and review meetings, even through they keep silent throughout; Community members are actively involved in decisions about how to implement a particular project and provide some of the necessary support and inputs.

2. Feedback Model

This Model has become popular with the growth of FSR and incorporation of on farm experimentation into agricultural research methodology. On the extension side, the T and V System is an example of the feedback Model.

Salient Features

- ☆ Researcher values identification of target group, his problems and reactions.
- ☆ Close interaction between researcher and extension system.
- ☆ Researchers are carried out both on research farm as well as farmers' fields.
- ☆ Team of researchers included extension personnel as well as social scientists to have a holistic understanding of farmers' problems.

3. Participatory Technology Development (PTD)

PTD is a process of bringing together the local communities with their knowledge and capacity for research on one platform with the researchers and other professionals.

PTD refers to enabling the farmers to develop and manage their own technology.

Features

- ☆ People centred
- ☆ Need based
- ☆ Site specific
- ☆ Holistic
- ☆ Farmer research based
- ☆ Quicker diffusion.

4. Indigenous Technological Knowledge (ITK)

Traditional Knowledge (TK) is gathered over a period of time and even passed on from generation to generation by words of mouth. It is highly restricted and localized.

5. Farming System Research/Extension (FSR/E)

Faming system through adoption of farm management techniques gave reasons to believe that multi- disciplinary teams incorporating social scientists can make important contribution to understand farming system research/extension (FSR/E) evolved asd a response to the need to identify opportunities for appropriate technology change among poor farmers (Farrington and Martin, 1998).

Characteristic Features of FSR/E

- ☆ It is strongly applied and empirical problem solving approach and is aimed at generating viable technologies in a shorter time.
- ☆ Mainly concerned with 'bottom up' research strategy, which begins with an understanding of the existing farming system and the identification of key constraints.
- ☆ Assessment of scope for and potential impact of technology change within a holistic farming systems framework.
- ☆ Identification of relatively homogenous groups of(usually resource poor) farmers within specific agro- climatic zones as the client of research and extension.

- ☆ Dynamic interactive approach in which one year's trials results generate hypotheses for testing retesting in the next.
- ☆ Concerns that the results of farm trials should influence on station research priorities.
- ☆ Effective linkage of researchers and extension workers (Okigbo, 1989).

Sequence/Activities Involved in FSR/E

1. Target and research area identification.
2. Problem identification and development of a research base through quick planning on farm research.
3. Planning on the farm research
4. Analysis of results of on farm research.
5. Dissemination of results through multiplication testing; and
6. Effective collaboration with other closely associated with the needs of resources poor farmers.
7. Models associated with FSR/E:
   a. Farmer- Back to Farmer Model (Rhodes and Booth, 1982)
   b. Farmer- First- and Last –Model (Chambers and Ghidyal, 1985).

Conclusion

The public extension system needs change for better services by structural reforms and improving function through transferring the ownership to and sharing responsibility of its management with community and by building synergic relationship between community and the public extension system. Communitisation would strengthen the role of public sector and it should facilitate introduction for common regulations against the particular interests of individual members. Communitisation would strengthen the role of public sector and it would facilitate introduction of common regulations against the particular interests of individual members. Communitisation of public extension system can be achieved through participation of the community. Similarly, other models and approaches suggested in the present paper seem to be befitting in today's context.

References

Chambers, R. and Ghildyal, (1984). Agricultural research for Resource poor farmers. The Farmer- First- and last- Model. Paper presented in the workshopon NARP of NAARM, Hyderabad (A.P.).

Chambers, R., Pacey, A., and Thrupp, L. A. (1989). Farmer first: Farmer innovation and agricultural research. London: Intermediate Technology Publications.

Farringtone, J. and martin,.A. (1998). Farmer perception in Agricultural Research. A review of Concepts and priorities, Nottingham; ODI.

Okigbo, B.N.(1989). The Farming system approach and links Between Research and Extension, In: Agricultural Extension in Africa, Nigel, R.(ed), World bank, Washington.

Rhodes, R.E. and Booth, R.H. (1982). "farmer- bach-to farmer". A Model for generating Acceptable Agricultural Technology. Agricultural Administration, II, pp. 127-131.

Chapter 3
Emerging Role of KVKs for Women Empowerment

U.S. Gautam[1] and Alka Singh[2]

[1]Director, ICAR-Agriculture Technology Application Research Institute (ATARI) Zone IV, Kanpur, U.P.
[2]JNKVV-Krishi Vigyan Kendra, Sidhi, M.P.

Women are key players in agriculture and allied fields. Rural women play key role by working with full passion in production of crops right from the soil preparation till post harvest activities (Ahmed and Hussain 2004). Their activities typically include producing agricultural crops, tending animals, processing and preparing food, working for wages in agricultural or other rural enterprises, collecting fuel and water, caring for family members and maintaining their homes (SOFA Team and Cheryl Doss 2011). The nature and extent of women's involvement in agriculture, no doubt, varies greatly from region to region. Even within a region, their involvement varies widely among different ecological sub-zones, farming systems, castes, classes and stages in the family cycle. Women make a significant contribution to food production and to the processing and marketing of foodstuffs. Despite their tremendous contribution in agriculture and allied fields, her hard work has mostly been unpaid. Role of women in agriculture is hugely emphasized in KVK. One of the principal thrust area of KVK is to make farm women participation more in agriculture and allied aspects. KVK is exploring the possibility of entrepreneurship development for rural women through agriculture.

Mandate of KVK

Assessment, refinement and demonstration of technology/products

Activities of KVK

- ☆ On farm testing to identify the location specific technologies in various farming systems
- ☆ Frontline demonstrations to establish production potentials of newly released technologies on farmers' fields and provide feedback
- ☆ Training of farmers and farmwomen to update their knowledge and skills in modern agricultural technologies and training of extension personnel to orient them in the frontier areas of technology development
- ☆ Work as knowledge and resource centre of agricultural technology for supporting initiatives of public, private and voluntary sector for improving the agricultural economy of the district
- ☆ Create awareness about frontier technologies through various extension activities like Farmer fair, Field day, Strategic campaign, Ex-trainees meet, etc.
- ☆ The seed and planting materials produce by the KVKs also be made available to the farmers

To identify the problems of farm women and increase the rate of their participation in different programmes and to make out the emerging issues several workshops were organized to sensitize the women scientists working in KVKs of Zone VII for rural upliftment and overcome their problems through development of modules for women in agriculture.

Objectives of the Workshop

- ☆ Participation in agricultural and economic activities,
- ☆ To identify factors that hinder their development,
- ☆ To overcome the drudgery and enhance efficiency,

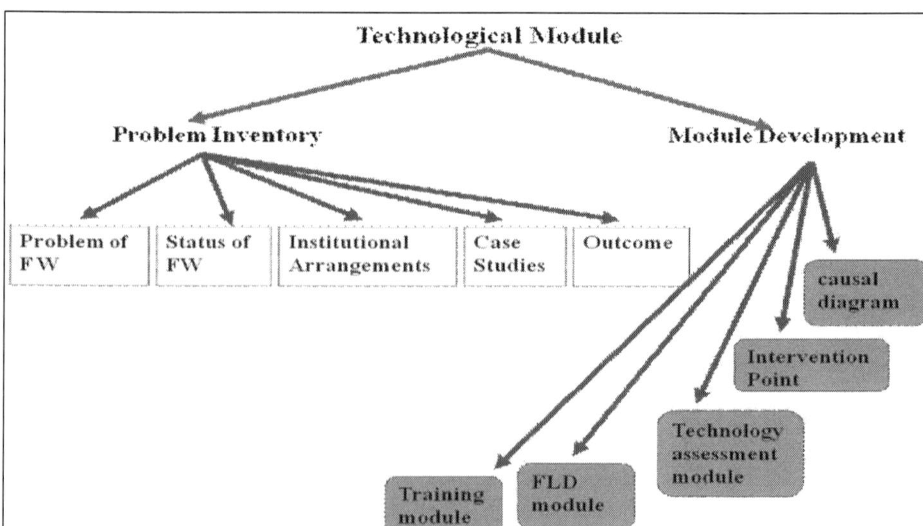

- ✰ To encourage for utilization of available resources for livelihood
- ✰ To promote them as an entrepreneur.

The outcome of the workshop came out in shape of a Technological module for farm women which contains problems of farm women, status of farm women, Institutional Arrangements, Case Studies, Outcome, problem inventory and process for the development of modules in terms of technology assessment, refinement and demonstration of technology/product and capacity building of farm women, which ultimately lead the improvement of livelihood of farm women.

On the basis of the modules, we identified 124 problems of farm women which were related to the priority thrust areas as income generation, drudgery reduction, value addition and nutritional security. The total 48 problems identified by Home Scientists of Madhya Pradesh were compiled in a book which is "Technological Modules for farm women of Madhya Pradesh", 36 problems of Chhattisgarh in "Inventory of Agricultural Technologies for Chhattisgarh" and 40 problems of Orissa were compiled in "Agricultural Technologies modules for Orissa". These books serve as a guide for all the Home Scientists to work properly in their fields and achieve their goal as upliftment of economic and social status of farm women.

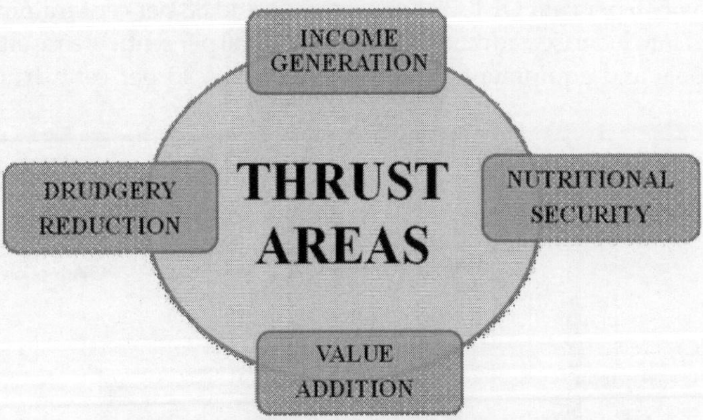

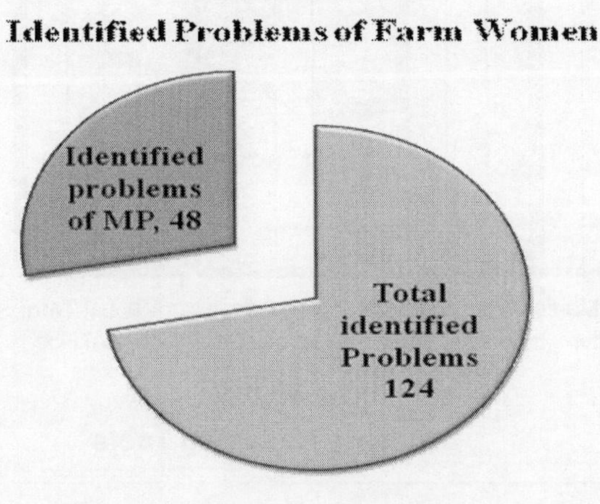

Result

From the year 2007-2013-14, KVKs conducted total 3490 trials on given thematic areas, a total no. of 4582 demonstrations were conducted and 6671 trainings were organized for empowerment of farm women.

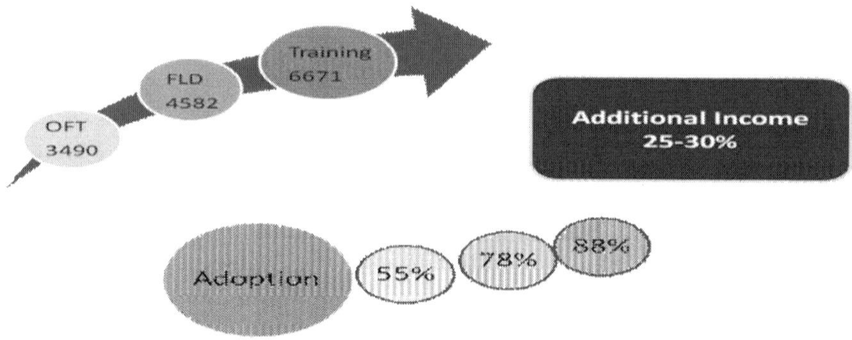

On the basis of developed module, KVKs worked in their areas according to the identified problems of farm women from 2007-08, All these interventions were resulted in adoption of 55 per cent OFT, 78 per cent FLD, and 88 per cent trainings which in turn resulted into increased additional income by 25-30 per cent. As a result of drudgery reducing tools and equipments Maize Sheller saves 43 per cent drudgery while

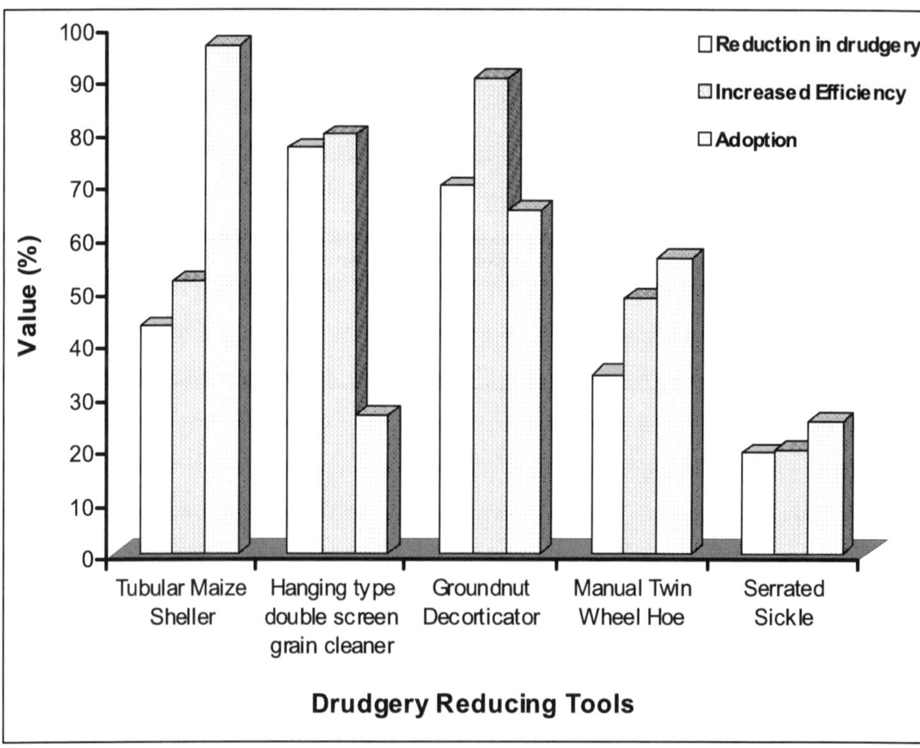

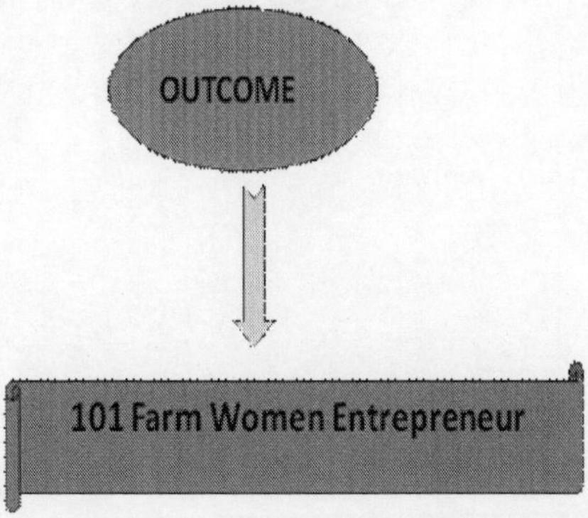

hanging sieve, groundnut decorticator, manual twin wheel hoe and Serrated Sickles are reducing 77, 70, 34 and 20 per cent drudgery of farm women, saves more than half the time and increases 52, 79, 90, 48 and 21 per cent working efficiency of farm women. After technology demonstration, the adoption of Maize sheller is 96 per cent, hanging sieve is 26 per cent, groundnut decorticator is 65 per cent, manual twin wheel hoe is 56 per cent and serrated sickle is 25 per cent.

Apart from these and after the above said activities we found 101 farm women entrepreneurs. These activities motivated them to stand on their own legs. They not only created jobs for themselves but for others also. They have taken up materials available in their areas for enterprises. They generated income through mushroom production, vermicomposting, rearing of birds, nutrition gardening, Growing of organic vegetables, flowers, oilseeds, seed production, bee-keeping, value addition, processing, packaging, dehydration of fruits and vegetables, canning or bottling of pickles, chutneys, jams, squashes, dairy and other products, stitching of clothes, knitting, weaving, embroidery, Bakery and flour milling etc. They have become socially and financially self sufficient and brought extra income for their families.

Conclusion

Socio-economic empowerment has been challenging task for the farmwomen. All these interventions were resulted in increased additional income by 25-30 per cent and 101 farm women stands on their own leg and they are now called as entrepreneurs. Therefore, this module proved worthy in improving the socio-economic status of the farm women in the region. It serve as a guide for all the home scientists by which they can choose an intervention and activities to be taken up in their respective places as per the problem identified. KVKs are not only building rapport with farmers and farm women but improving and increasing their socio- economic standard of living and sources of revenues.

References

Ahmed N., Hussain A. (2004). Women's Role in Forestry: Pakistan Agriculture. Agriculture Foundation of Pakistan, Islamabad, pp. 79-81.

SOFA Team and Cheryl Doss (2011). The role of women in agriculture. ESA Working Paper No. 11-02. Food and Agric. Org. Agric. Devel. Econ. Division, P. 47.

Chapter 4

Linkages between Research and Extension for Technology Transfer

B.U. Dupare[1] and Prashant S. Armorikar[2]

[1]ICAR- Indian Institute of Soybean Research, Indore, M.P.

[2]Directorate of Extension, Department of Agriculture and Cooperation, Ministry of Agriculture, New Delhi

1. Introduction

Agricultural research findings may lose its validity, if they are not adopted by the farmers appropriately. Unfortunately, the only effect of research is to produce an interesting article for finding a place in a scientific journal or report without its utility. Hence, research must be related to farmers' problems if we are to avoid this situation (Ban, A.W. Van den, 1994). The lack of a close working relationship between national agricultural research and extension organizations, and with different categories of farmers and farm organizations, is one of the most difficult institutional problems confronting ministries and departments in many developing nations. To understand the different roles that research, extension, input supply dealers, and other system actors should play in the technology development and transfer process, it is essential to differentiate between the different categories of technology. Each category of technology follows a different channel as new technology is developed and transferred to farmers for adoption (Swanson, 1993).

Most of the research systems in developing countries were focused to serve commercial farmers operating in more favorable and homogenous agro-ecological conditions than those in resource-poor farming contexts. Traditionally, resource-poor farmers have lacked the potential and organizational ability to exert pressure on the research system to meet their needs. Their access to research information and capacity to tolerate risk is limited, and the pressing concerns of their daily existence make it difficult for them to focus on long-term technological change (Merril-Sands *et al.*, 1989). Most of the developing countries concern about rural poverty and malnutrition, unsustainable agricultural production, and swelling of urban population. Part of the problem is due to lack of effectiveness of agricultural research at farmers' field (Eicher, 1988; Jain, 1992).

The process of transfer of technology (TOT) may not be relevant, if applied in isolation. It is long and involves various institutions and stakeholders at different stages from research stations to the farmers' field. The effectiveness of TOT is greatly influenced by the linkages between and among those partners which are responsible for technology generation (research sub-system); technology transfer (extension sub-system); those who actually put to use the technology (client sub-system). Thus TOT of any improved practice calls for concerted and coordinated efforts and mutual understanding between and among the above sub-systems involved (Kunju, A.R., 1992).Before going into details about the sub-systems in TOT in agriculture, it is important to know some of the key concepts and definitions which are used in this paper. Besides, the issues concerning non-effectiveness of technology generation and transfer at field level are also required to be incorporated in this document.

II. Concepts

(i) Agriculture Knowledge Information System (AKIS)

An AKIS is defined as a set of agricultural organizations and/or persons, and the links and interaction between them, that are engaged in such processes as the generation, transmission, storage, retrieval, integration, diffusion and utilization of knowledge and information with the purpose of working synergically to support decision making, problem solving and innovation in a given country's agriculture or domain thereof (Roling, 1990). The AKIS, in addition to its actors, has many other stakeholders, such as processors, merchants, consumers, policy makers and donors who have vested interest in it and can indirectly influence it.

(ii) Research

Research covers the whole spectrum of scientific enquiry and technology development, from basic to adaptive research. The main tasks of research are:

- ✰ Discovery of innovative technologies
- ✰ Exploratory development
- ✰ Technology consolidation

(iii) Technology Transfer

'Technology transfer' refers to the process of (a) bringing research results, in the form of new agricultural technologies and new information, to farmers and (b) supplying research with information on farmers' needs, production constraints and feedback on technologies. The main tasks of technology transfer are:

- delivery of technologies to farmers in appropriate manner
- continuous review, monitoring of technology users for further improvement

(iv) Technology Validation

To prove the effectiveness of technology innovation at field level.

(v) Policy Makers

These are in-charge of defining technology policies, approving research strategies and managing other development instruments such as pricing public investment and market regulation.

(vi) Research Extension (R-E) Linkage

Most important link between lab and land. It is important to include the role of inputs and services in the discussion of technology development and delivery."Extension" is now usually associated with conventional public sector extension services. "Technology transfer" however, can be applied not only to these services but also those provided by many other institutions/organizations *viz.* private firms, parastatals, NGO's, farmers organizations etc.

(vii) Linkage and their Mechanisms

Linkages are defined as the channels for the two way flow of knowledge, information, and resources between research and its partners. The links can be discussed as given below:

- **Functional links** - This relates to research and technology transfer activities.
- **Institutional links** - which relates to the institutions and personnel that carry out these activities, which aims to form a bridge between research and technology transfer.

The research or extension manager can provide oversight and coordination for each set of functions or linkages all the way down to individual researchers and SMSs who are carrying out their own programme of work in developing, testing, and disseminating a package of recommendations for a particular commodity or farming system.

The organizational procedures used to establish, maintain or improve links are termed as "linkage mechanisms" with following characteristics:

- Whether they are formal/informal, regular/ad hoc, mandated/voluntary, permanent/temporary
- Whether they are facilitative mechanisms
- The amount and type of resources exchanged
- Administrative level at which they operate
- Whether they focus on programming activities or are concerned with the implementation or evaluation.
- The numbers of individuals involved.

Linkage mechanisms are used to channel information between groups and to coordinate required tasks in the process of getting relevant technologies to farmers. In the process, these linkage activities help to improve resource use by avoiding the duplication of effort and ensuring that critical tasks do not fall through the institutional cracks (Merrill -Sands, 1992).

Basically, there are two types of linkage mechanisms (i) Organizational and (ii) Managerial.

(i) Organizational Linkage Mechanism

involve the structural modification of the research and / or extension organization or other organizations in the process. These modifications may range from *the formal merger of* research and extension at the broader system level, the merger of specific units within research or extension, or it could involve the creation of new positions, units, or permanent committees. In general, when reorganizing, be sure to combine those groups that are either dependent or that need to communicate with each other to get a job done. Also, put people together whose work might overlap to avoid duplication of activities (Merrill-Sands, 1992).

(ii) Managerial Linkage Mechanism

The other major type of linkage mechanisms involves a range of managerial interventions. For example, research and extension may agree to collaborate on joint planning and review activities, such as conducting RRA or PRA in different agro-ecological zones to assess the farming systems and technology needs of different groups of farmers. These activities would likely result in joint priority-setting and in planning joint programme activities (*e.g.*, On-Farm Trials and Demonstrations). A second category of managerial linkages is when researchers and SMSs actually carry out *collaborative* programme activities together, such as on-farm trials and demonstrations, plus joint decision making on technical recommendations. In the process of carrying out these different programmatic activities, researchers and their extension counterparts share the responsibility for different tasks and regularly consult each other on informal basis.

In addition to these formal linkage mechanisms, both research and extension personnel would be expected to have regular, informal contacts with different groups of farmers in their respective service area.

Eponon (1993) cited by Crowder and Aderson (1997), stated six broad functions for research –extension linkage, these are:

- ☆ Technology planning and review;
- ☆ Collaborative activities *e.g.* field trials that combine the experts of both researchers and extensionists.
- ☆ Exchange of financial and human resources *e.g.* staff secondment;
- ☆ Exchange of knowledge and information through joint reports, field visits, training etc;
- ☆ Feedback on the performance of technology through evaluation, surveys and field visits;

- ☆ Coordination of information flow between institutions or units using for example research extension liaison.

The system of research–extension linkage is an important factor influencing the success of an extension organization **(Agbamu *et al.*).** In several government extension organizations there is a well established system of subject-matter specialists who keep field workers informed about relevant new developments in research and inform researchers about the problems and experiences of farmers.

III. Need for Links

Strong links ensure that:

- ☆ Research tackles users priority needs and problems;
- ☆ Farmers and technology transfer workers keep up with research developments;
- ☆ Research results from experiment stations are applied to solving farmers' problems and their opportunities;
- ☆ Available technologies are adopted to suit local agro-ecological and socio-economic conditions;
- ☆ Successful technologies are promoted and distributed widely to farmers;
- ☆ Users have access to the information, inputs and services required to support a technology;
- ☆ Researchers can capitalize on users' knowledge and obtain feedback on the relevance and performance of technologies.

IV. Criteria for Evaluating Links

- ☆ **Integration of AKIS:** A high level coordination, collaboration and communication within AKIS is a pre-requisite for high system performance. The level of integration is gauged according to the amount of resources exchanged between the parts of AKIS and the importance that each part attaches to these resources.
- ☆ **Relevance of new technologies:** A new technology is relevant to a group of farmers if it responds to their needs. The best way to assess the relevance is to look at the extent of its adoption.
- ☆ **Responsiveness of new technologies to the needs of resource poor farmers:** It is same as relevance but emphasis is given on resource-poor farmers.
- ☆ **Institutional sustainability:** Evaluation criteria are the ability of an institution to sustain its performance over a period of time.

Factors Affecting Research Technology Transfer Links

Major Issues and Problems affecting the Integration of Agricultural Research and Extension are grouped under following categories:

- **Policy/political and structural considerations** (*e.g.*, lack of clear specification of organizational responsibilities and/or absence of unified policy framework
- **Resource factors** - inadequate incentives or government support (*i.e.*, lack of sufficient funds);
- **Technical factors** - compartmentalized structure of research and extension (*e.g.*, existence of institutional boundaries)
- **Organizational factors** - internal organizational pressures and cultures;
- **Inappropriate Handling of Research Technology** - lack of adequate access to quality information; and
- **Lack of efficient and effective strategies** - to strengthen REF linkages

(i) Policy/Political and Structural Considerations

The policy context is defined, to a large extent, by the external pressures on research institutions to improve performance, develop links, and address the needs of resource poor farmers. The external pressure has some positive as well as negative impact on the performance of an institution.

(a) Political Pressure

- **Historical legacy:** Foreign settlers and indigenous elites had close contact with researchers. Researchers are endeavored to meet commercial farmers' needs for their own benefits. Whereas subsistence sector gets attention only where a particular commodity forms the basis of national economy or in exceptional cases such as crop failure and famine, disease outbreak, rapidly rising prices or imports etc.
- **Current political and social structure:** For political reasons, governments are reluctant to allow farmers organizations to be formed or become too powerful. The only type of organizations found among resource-poor farmers are informal networks of *patron-client* type, getting most of the benefit in terms of credit, inputs or services by the patron only.

(b) Effect of External Pressure

Good institutional performance requires positive pressures on AKIS by national policy makers, foreign donors, farmers and private sectors. It motivates institutions to develop and deliver relevant technologies to resource-poor farmers.

(c) Without Pressure

- Research will become less applied
- Technology transfer agents will become less applied
- Researchers and technology transfer agents will devote more time to meet the demands of richer farmers
- Links between *on-station* and *on-farm* and between researchers and technology transfer agents

(d) Limitations of External Pressure

- ✰ External groups' interest may fluctuate or be short term, whereas technology development and delivery is a long-term process.
- ✰ Institutions may try to create the impression that they are responding to external demands without really doing so; this makes underlying problems more difficult to identify and solve.
- ✰ Institutions may compete for support from external sources; this can damage effective collaboration between them.
- ✰ Demands made by external sources may conflict with each other, pulling an institution in several directions and reducing overall performance.
- ✰ Pressure from external sources, particularly foreign donors, may not reflect national priorities.
- ✰ Demands made on institutions may not take sufficient account of the capacity of institutions to meet these demands (Merril-Sands *et al.*, 1989)

Resource Factors

For sustaining effective research – technology transfer – user links, financial and human resources must be allocated to support linkage activities, however, most developing countries are facing the problems of resources.

(a) Linkage Problems Arising from Resource Constraints

- ✰ Inadequate funds promote competition between institutions rather than cooperation, for seeking as much from limited resources.
- ✰ Resource scarcity often makes institutions exert tighter control.
- ✰ While introducing cutbacks in expenditures, linkages are viewed as desirable but not essential.

(b) Overcoming Resource – Related Linkage Problems

- ✰ Linkage strategies should be limited in scope and fairly simple to implement.
- ✰ More networking should be undertaken, using farmers, para-professionals, rural development projects and NGO's to carry out technology testing and dissemination.
- ✰ It may be desirable to recruit ideologically motivated staff, willing to work in remote areas and accept lower salaries.

(ii) Technical Factors

(a) The Knowledge Base

The knowledge base is defined by how many different actors – scientists, technology transfer agents, and farmers – know about the environment and the relevant technologies. Scientists build up knowledge through research; farmers and technology transfer agent acquire it through experience and from other farmers. By involving farmers directly, research conducted on-farm offers a means of expanding the knowledge base and providing useful feedback. So close interaction between them will strengthen base.

(b) Agro-ecological and Production System Diversity

Most developing countries are characterized by diverse environments in which resource-poor farmers, using a wide variety of production systems, are striving to produce a wide range of agricultural commodities. Diverse environment require more location specific diagnosis and more technology adaptation. The resources needed to cope with diversity are lacking. Diversify makes it more difficult to have one simple message. This creates extension ineffective in these areas.

(c) Communication and Input Distribution Infrastructure

Many communication channels can be used but the choice of which in a given situation will depend upon farmers' access to these channels and their ability to use them. Illiteracy among farmers is a severe constraint which most developing countries are facing. Where public distribution systems, particularly in private sector, are weak, extension services often concentrate on input delivery. Dissemination of technical information becomes less important part of their work, reducing the links with research.

(d) Dispersed Farming Populations

It makes fewer opportunities for informal interaction. This increases the need for formal linkage mechanisms to facilitate interaction. To be effective, there must be some measure of decentralization.

(iii) Organizational Factors

The degree to which a system is integrated depends partly on how it is organized. Some of the issues to be considered while integrating research and technology transfer are given below.

(a) Interdependence

The degree to which tasks are seen as interdependent also influences the way in which they are grouped in organizational units. Earlier, research and technology transfer/extension were considered to be independent of each other but contributing to a common purpose. Later, extension was considered to be sequentially linked to research, receiving its input from research and incorporating these into a package of services to the farmer. Subsequently, their reciprocal interdependence was recognized, with extension indentifying problems and supplying information which enabled researchers to define priorities. This change of concept in the literature has not, however, been fully accepted by the relevant institutions themselves. Today, we need a team approach with a strong two-way flow of information and an attitudinal change between these partners.

(b) Merging Research and Technology Transfer

It is very frequently recommended in literature as a way of increasing integration. However, in practice it has some problems. Two major advantages to merging are usually cited. First, organizational proximity can promote a shared goal and facilitate communication and collaboration researchers and technology transfer workers.

Secondly it will increase the efficiency. On the contrary, merging frequently raises unexpected problems (Bourgeois, 1989). Merging two groups in one institution or unit is no guarantee that they will work together eg. Gambia. The second problem is that, quality of research can be jeopardized if it becomes 'submerged' in the development process. By merging, the combined institutions' resulting size will be unmanageably large. The loss of autonomy can lead to conflicts and growing resistance among personnel who see their independence increasingly threatened. Merging will be useful only where a system is organized around a specific region or, commodity.

(c) The differences between Researchers and Technology Transfer Workers

For high performance, specific linkage mechanisms are required to manage the conflicts and communication problems caused by differences between researchers and technology transfer workers in background, training, experience, responsibilities, status and physical location.

- ☆ **Informal groups**: It makes communication and collaboration easier. The differences between researchers and technology transfer workers have major implications for communication. According to communication research, heterophilly reduces communication.
- ☆ **Occupational groups:** Such as researchers or extension agents, compete with each other for status and rewards. This makes rival groups, on the basis of academic qualifications (Bennel, 1989). Thus, to justify own status, researchers distance themselves from lower status. These strong status differences between occupational groups are difficult to bridge through linkage mechanisms.

(d) Personnel and Financial Management

Personnel and financial management policies and practices which encourage integration and provide flexibility in AKIS result in higher level of performance. Differences in policies and practices between research and technology transfer institutions can greatly hinder the integration of two activities. The candidates for technology transfer may not include people with right qualification and skill. It is hard to attract staff to some geographical areas. Job descriptions are often vague, non-existent; little emphasis is given on collaborative activities. High levels of graduate unemployment create pressure to hire large number of staff who cannot be effectively used. All these hampers the institutions ability to develop effective relations with others groups.

(e) Size of Institutions

Larger institutions are more complex, tasks are divided among a greater number of specialist groups, and scientists tend to be more isolated from users. Hence multiple linkage mechanisms of more formal nature are required.

(f) Reorganization

It is one of the interventions considered by managers. However, it requires high cost; merely changing organization does not bring about the operational changes required in the field. Some changes may be needed from time to time to accommodate changing circumstances, but major one should be undertaken with the greatest care. Senegal reorganized its system with the primary objective of strengthening links with farmers and other agencies weakened its links further after formation of a production system department (Faye and Bingen, 1989).

International Service for National Agricultural Research (ISNAR) in its programme started to work with farmers' organizations in three African countries *i.e.* Burkina Faso, Ghana and Kenya. The five types of farmers' organization they found were traditional associations; informal contact groups; multipurpose associations; cooperatives and national farmers' organizations. ISNAR further reported the reasons, why most of these organizations do not have linkage policies. These were:

- ☆ Access to technology is not one of the objectives of organizations.
- ☆ The organizations were not convinced of benefit from linkages.
- ☆ Organizations are unaware that they can establish linkage with research.

However, those organizations, having linkage problems faces the management problems because:

- ☆ Many researchers are not fully convinced of usefulness of linkages.
- ☆ Linkages are perceived as extra activities by researchers.
- ☆ Research policies, like rewards and promotion are not compatible with having strong linkages with farmers.

However, above problems can be minimized by:

- ☆ Balancing the decision making power between research and farmers organizations with regards to setting and implementing research agenda.
- ☆ Jointly defining sound policies and strategies (Eponou, 1996).

(iv) Appropriate Handing of Research Technology at Field Level

- ☆ It is evident that despite effective technology generated by the scientists, the results at farmer's field level may not be desirable due to its mishandling at its various stages of implementation.
- ☆ It is also important to develop crude technology vis-à-vis indigenous technology know-how so that the farmers would handle at their own level.

The high participation of extension agents in linkage activities implies that when extension is the focus and mandate, resources are likely to be utilized effectively in carrying out extension functions (Oladele, 2001). Farmers' non-involvement in many other activities underscores the lack of complete or partial linkage existing between researchers, extension agents and farmers. While it is a known fact that the majority of farmers are illiterate, they have been found to be repertoires of rich indigenous knowledge that are sources of vital information (local), which will enhance the development, acceptability and utilization of technologies.

A Faturoti *et al.* (2007) reported that a fragile and poorly defined relationship was observed among stakeholders in the plantation and banana innovation system, and concluded that these need to be strengthened.

Available evidence shows that yields of major crops under farmers' management are still far lower than what can be obtained under research managed plots (Abate, 2006; EIAR, 2007). In this regard, Ahmed, Kelemework and Abate (2006), noted that under Ethiopian conditions, the potential yields of improved varieties of haricot bean, durum wheat, bread wheat, finger millet, and sorghum was much lower. This is a clear indication of the gap, which exists between researchers and farmers. The absence of effective linkage between agricultural research and extension systems has repeatedly been reported as one of the major reasons for the low productivity of Ethiopian agriculture.

Different types of technologies require different linkages mechanism. Lancini (1987) reported that, for an upland cropping project in Philippines, they brought together the researchers and technology transfer workers, and diagnostic model was developed for solving serious problems. Again steering committees were formed representing both disciplines for planning, coordination and monitoring. This ensured close relationship between the disciplines. Doorman (1991) emphasized on need to involve farmers in setting research policies with analysis of production constraints at farm level, while studying the case of rice in Dominican Republic.

VI. Factors Responsible for Improving Agricultural Research and Extension Linkages

- ✰ Viewing research and extension as an integrated knowledge and technology system;
- ✰ Focusing on research and extension functions, rather than simply on organizational structures, to broaden dialogues among all stakeholders;
- ✰ Strengthening crucial linkages under a pluralistic approach/system;
- ✰ Promoting informal networking at many levels with an incentive system for improving linkages; and
- ✰ Empowering farmers' groups for more demand- and farmer- driven research and extension services.

VII. Role of ICT in R-E-F Linkage

While ICTs offer opportunities for effective agricultural information and knowledge management, it remains an open question whether it pays off to invest resources in establishing such systems and how the systems should be designed and managed to keep them relevant and practical. The new ICTs offer opportunities for more effective agricultural information and knowledge management. However, while such integrated systems could certainly facilitate access to information, their usefulness will depend on continuous up-dating and quality control. Both, building up and maintaining knowledge banks uses up considerable resources. It is therefore a question whether it really should be a priority to invest major efforts into building such systems, and how such systems have to look and be managed in order to avoid that they become further information dumps. In the case projects generally research actors are looking at research outcomes and extension at extension outcomes. Sometimes, research relies on extension for data collection (Bhutan, IRRC). To reach farmers on such a large scale in an interactive mode, the need to harness the projecting of ICT becomes imperative. IT is the only mechanism which has the potential to reach out with technologies, knowledge and information extensively and rapidly. It is also the only instrument that can keep pace with information regarding changing market prices, arrivals in markets, global market scenario, weather, etc.

VIII. Convergence between Agricultural Research and Extension

The Department of Agriculture and Cooperation (DAC) has been mainly involved in designing programmes and schemes targeting specific areas in agriculture sector, the Indian Council of Agricultural Research (ICAR) has been providing the necessary scientific research back-up for better scientific and technological inputs to improve productivity.

There is a scope for improvement in the interaction between the scientific community and the field functionaries who are responsible for implementation of schemes and programmes for the benefit of the farmers. One of the reasons for the success of the Green Revolution was that scientists, administrators and farmers all came together to engineer a paradigm shift in the way agriculture was practiced. Over the years, this cohesion has not been all that strong. More often than not the farmer is deprived of the benefits of scientific and technological advancements due to lack of interface between the scientist and the administrator, between the administrator and the farmers and between the scientist and the farmer. It is, therefore, important for the departments and the Institutes to rethink their strategy on how best they can bring about a direct interface among the scientist, the administrator and the farmer or in other words how successfully they roll out the "lab to land" approach.

In the States, the agricultural research and education system is available through the State Agriculture Universities (SAU), several specialized ICAR institutions and the KVKs. Similarly, the extension mechanism through ATMA (Agricultural Technology Management Agency) and the State Agriculture Officers and lower functionaries is available for extension authorities and for rolling out various schemes of Central and State Governments. The convergence mechanism states that;

- ✰ The Programme Coordinators of the KVKs in the District should have an interface meeting with the Project Director, ATMA regularly during the cropping season and work out a strategy of providing crop advisories to farmers for various stages of crop growth.
- ✰ Scientists in the KVKs shall advise and mentor Block Technology Teams in identifying technological needs in various Blocks in the District and programmatic interventions to meet such needs.
- ✰ SAUs need to designate an expert who would be in-charge of one District to provide instant guidance to extension functionaries in the District.

- ✰ District Agriculture Plan or District Agriculture Action Plan developed on the basis of the Strategic Research and Extension Plan (SREP) should be refined in the process jointly by ATMA and KVKs from the Block level and acted upon for the purpose.
- ✰ ATMA, KVKs and SAUs need to coordinate their efforts in research and extension. ATMA and KVK should coordinate with each other in the conduct of Field Days, Kisan Melas, Goshties and setting up Farm Schools, so that there is no duplication in coverage and they should ensure percolation of appropriate scientific practices down to the field level.
- ✰ The KVK and SAUs may provide an Agricultural Technology Update (ATU) before the start of Kharif and Rabi crop seasons for its wider dissemination among the farmers.
- ✰ Extensive use of ICT application may reduce the gap between technological research and adoption by the farmers at field level.

IX. Managing Linkage Mechanisms

Strong linkage among Research –Extension –Farmer (REF) can be considered as a precursor for undertaking and implementing any extension programme at field level effectively. Different types of technologies often require different kinds of mechanism. These differ for improving collaboration and communication. ISNAR studies have recommended following types of linkage mechanisms:

- ✰ Joint problem diagnosis – effective in field situation.
- ✰ Joint priority setting and planning exercise – 'Samuhik bhramans' in Nepal have been effective in developing link (Kayastha, *et al.*, 1989).
- ✰ Joint programming and review meetings – effective when small, concentrate on particular commodity.
- ✰ Formal collaboration in trials, surveys, and dissemination activities.
- ✰ Joint field visits – to see actual performance of recommendations.
- ✰ Joint decision making on the release of recommendation.
- ✰ Formal guidelines for allocating time to linkage activities – managers need to provide clear and realistic time allocation guidelines.
- ✰ Specific allocation of funds for linkage activities.
- ✰ Staff rotation – to pool experience and get collaborative activities moving.
- ✰ Training and seminars.

X. Suggestions for Better Research Extension Linkages

- ✰ Involvement of multiple actor – farmers, extension workers, ATMA, KVKs, NGOs, FAOs, Agri-preneurs, Cooperatives, SHEs, CIGs, Para-Professionals and input suppliers etc.
- ✰ Farmer – Scientists' interaction should be organized in the village itself for better linkages. At the same time scientists should get an opportunity to meet the extension personnel in farmers' field.

- ✰ Farmers' problems should reach the scientists well in time either directly or through agriculture supervisors so that they may be solved at the earliest.
- ✰ There should be more specialized staff at the field level to provide immediate solution to the farmers' problems.
- ✰ The presence of agriculture supervisor of the respective area should be made mandatory in any training or meeting organized for farmers.
- ✰ Farmers' fair should be a regular feature in rabi as well as kharif *i.e.* at an interval of six months to provide the farmers with latest know how and development.
- ✰ Emphasis should be given on quality of demonstrations, mini-kits and skill based training.

The impact of technology could be examined on the basis of increased farm income, reduced risk, resource-conservation and overall socio-economic development of farmers.

References

Abate, T. (2006) Focusing Agricultural Research to Address Development Needs: the Way I See it. In: Abate, T. (Ed.), Successes with Value Chain, Proceeding of Scalling up and Scaling out Agricultural Technologies in Ethiopia: an International Conference, 9-11 May 2006, Addis Ababa, Ethiopia, p.p. 1-20.

Agbamu, J.U. (2000). Agricultural research extension linkage systems: andinternational perspective. Agriculture Research and Extension Network. No. 106, pp. 1-7.

Ban, A.W. Van Den and H.S. Hawkins (1994). Agricultural extension. Longman scientific and technical co. New York pp. 26-33.

Bennel, P. (1989). Intergroup relationship in institutional agricultural technology systems. In making the links: Agricultural technology transfer of developing countries, edited by D. Kaimowitz, The Hague, ISNAR.

Bourgeois, R. (1989). Promoting integration through structural changes: Making the link between research and technology users. ISNAR working paper.

Eponou, T. (1996). Partners in technology generation and transfer: Linkages between research and farmers organizations in three selected African countries. ISNAR Research Report No.9. The Hague, ISNAR.

Faturoti, B., Madukwe, M., Tenkouano, A. and Agwu, A.2007. A review of policy acts and initiatives in plantain and banana innovation systems in Nigeria.Afr.J.Biotechn.

ISNAR, (1987). Working to strengthen national agricultural research systems – ISNAR and its strategy. The Hague, Netherland.

Kaimowitz, D.K. (ed) (1990). Making the link: Agricultural research and technology transfer in developing countries. Westview press London (in cooperation with ISNAR).

Krall S *et al.*, 2003. Nature of Research Impact Assessment and Evaluation in Agricultural Research for Development.

Kunju, O.A.R. (1992). Transfer of agricultural technology: structural and functional linkages. Concept publishing co. New Delhi.

Lancini, J.B. (1987). Strengthening the research/extension linkage through appropriate technology: A case study from southern Philippines. Agric. Admn. and Extension Vol. 24: 61-67.

Merril – Sands, D., Kaimowitz, D., Kay Sayce and Simon Chater (1990). Technology triangle: Linking farmers, technology transfer agents and agricultural researchers Summary Report of an International Workshop held at ISNAR from 20th – 25th Nov., 1990.

Rolling, N. (1990). The agricultural research – technology transfer interface: A knowledge system perspective. In making the link edited by D. Kaimowitz, The Hague, ISNAR.

Swanson, B.E. (1993). Identifying linkage problems using systems analysis: A training module. Urbana, IL: International Program for Agricultural Knowledge Systems.

Chapter 5
Producer Collectives for Value Chain Intervention and Linkage with Market

R.N. Padaria and V.K. Sajesh

Division of Agricultural Extension,
ICAR-IARI, New Delhi

Introduction

Even with the best possible technologies, farmers alone cannot sustain the market pressures and imperatives. Linking with market and making farming remunerative cannot be assured only with individual's knowledge and skill rather it should be grounded in community –based actions. Formation of grassroots institutions like farmers' interest groups, commodity interest groups, collectives or cooperatives has emerged as the basic pre-requisite for effectiveness of market-led development. Though the efforts have been afoot to ensure participation of farmers in developmental process, more emphasis has to be laid upon structural arrangements where-in the farmers could be organized and enabled for collective action, quality production and value addition, taking initiatives towards agribusiness, and sensing and capitalizing opportunities for market integration.

Value chain development has shown immense potential for improving farm productivity and income through collective action of farmers. A value chain is the full range of activities which are required to bring a product or service from conception through the different phases of production (involving a combination of physical transformation and the input of various producer services), delivery to final customers, and final disposal after use (Kaplinsky and Morris, 2000). Major goal of value chain development is concerned with modes of linking producers to market.

Small farmers are confronted with number of constraints in different phases of value chain *viz.* in access to inputs, credit, technology, marketing and value addition. V.S.Vyas (2011) observed that small farm-based agriculture will grow to the extent the institutions of research, extension, marketing and credit are attuned to the requirement and capabilities of the small holdings. These institutions were at present largely bypassing these sections because of higher transaction costs, perception of high risks and low absorption capacity of small farms. The institutions supporting agriculture would become more effective if they introduce suitable organizational changes and use technology to reduce transaction costs. Output market is an important concern, where small holders are often confronted with price volatility and exploitative nature of market intermediaries. The marketable surplus of individual smallholders is too small to be traded economically in distant urban markets owing to high transportation and transaction costs (Birthal and Joshi, 2007). The low volume of marketable surplus also constrains the bargaining power of small holders. Analysis of the constraints of small holders point out that many of the small holders' priorities will only be met if strong institutions are in place to provide a strong support to rural population to reach markets, access knowledge, information and technologies and to influence politics that affect their lives as well as to overcome coordination failure in delivery of services.

It can be comprehended from the analysis of the constraints of small holders that many of these hurdles could be crossed if aggregation of various functions of small holdings takes place. For that strong institutions should be in place to provide support to rural population to access knowledge, information, technologies and markets. Many institutional forms and models evolved over time in this milieu, of which producers' collectives like cooperatives, self help groups, producer companies, commodity interest groups etc. assume significance.

The concept of collective action is re-emerging recently to organize small farmers in developing countries along value chain in the wake of agricultural market liberalization (Coulter *et al.*, 1999). Collective actions can improve farmers' access to resources such as inputs, credit, training, logistics and information and also spread the risk among farmers (Agarwal, 2010). It can increase bargaining power (Quizon and Riguer,2003), offer an alternative to state and markets institutions(Pandolfelli *et al.*,2007), allows smallholder farmers to jointly invest in management practices that provide collective benefits in terms of economic and sustainability gains (Shiferaw *et al.*,2008) and can help the smallholders to overcome the problems of transaction cost arising from their small size of marketable surplus (Kruijssen *et al.*,2009), Fischer and Quaim (2010) found that marketing through the group yields a higher price than selling individually at the farm gate. Producers' collectives like cooperatives, producer companies and other organizations offer a new avenue for the smallholder producers by facilitating linkages with various actors in agricultural innovation system and thereby providing access to hitherto constrained resources and services.

Cooperatives

Potential of cooperatives in empowering the small holders has been well demonstrated in the dairy and sugar sector. Cooperatives were initiated in these sectors as a response to the economies of scale required in processing and marketing as well as to avoid the exploitation by intermediaries. Dairy cooperatives account for the major share of processed liquid milk marketed in the country. About 60 per cent of the installed processing capacity for milk is in the co-operative sector. Milk is processed and marketed by 170 Milk Producers' Cooperative Unions, which federate into 15 State Cooperative Milk Marketing Federations. There are almost 140,000 village societies with 14 million farmers as members (NDDB, 2011). Most of the sugar production in India takes place in the mills owned by local sugar cooperatives. Small holders are given the status of share holders in the cooperatives and provide credit, inputs, services and information. In dairy and sugar sector cooperatives have their own marketing arrangements. Two important cooperatives in horticulture are Mahagrapes in Maharashtra and HOPCOMS (Horticulture Produce Cooperative Marketing and Processing Society Limited) in Karnataka.

Mahagrapes

Mahagrapes is an association of 16 grape growers' cooperative societies. Mahagrapes was established in the year 1991 as a marketing partner to the group of co-operatives in Maharashtra state. It owes its origin to the Maharashtra State Agricultural Marketing Board. Mahagrapes is the first organization in the state to have the characteristics of both a cooperative and a private sector partnership firm. Since the establishment it has been able to organize the efforts of grape growers spread across Nasik, Sanghli, Sholapur, Pune, Ahmednagar, Osmanabad and Latur. Objectives of Mahagrapes are elimination of middlemen in marketing process, encouragement and development of agricultural export, and the provision of extension services and inputs to the members (Mahagrapes, 2015). To fulfill these objectives it negotiates better price for its members and also provide technical assistance, inputs and information to the farmers to enable them in meeting stringent quality requirement of export markets. Mahagrapes performs different functions *viz.* to source and develop worldwide markets, to provide quality control in post harvest activities, supply and supervise branded packaging (Narrod *et al.*,2007). Mahagrapes supplies inputs, technology, and extension services to farmers through cooperatives and support the farmers to produce quality output conforming to food safety and quality standards of the importing countries. It provides refrigerated transport and cold storage facilities for farmers by charging a fixed amount (Birthal *et al.*, 2007).

The data from beneficiary farmers and non-beneficiary farmers of *MAHAGRAPES* from Sangali district revealed that an average additional profit of $ 3958 per hectare was secured by the beneficiary farmers as compared to non-beneficiary farmers. Adoption of proper production management practices; harvesting and post-harvest handling (grading and packaging) helped the members in securing higher price ($ 5.07 per box) than the non-members ($ 3.96 per box). High adoption behavior (79 per cent) and high level of satisfaction (50 per cent) were observed among beneficiary farmers. Similarly, in case of *MAHAANAAR*, which is an association of pomegranate growers, the benefit cost ratio for the members (3.25:1) was higher than the non-members (2.25:1). Hence, there is need to promote and replicate commodity based farmers' cooperatives model (like *MAHAGRAPES*) in other part of the country for the benefit of the farming community.

HOPCOMS

Marketing of highly perishable commodities like fruit and vegetables has always been fraught with problems. In Karnataka, farmers were in the clutches of the middlemen and the system benefited neither the farmers nor the consumers. Grape Growers Marketing and Processing Society was formed in 1959 under the Indian Co-Operative Society Act in order to tackle these issues. The members comprise farmers, state financial organizations and the Karnataka State Government (Kolady, *et al.*, 1997). Since grape was a seasonal fruit, the society started handling all types of fruits and vegetables from 1965. Due to this change in operations, the name of the society was changed into Horticulture Producers Co-operative Marketing and Processing Society Ltd. From 10th Sep 1959 (Rao, 1997).

Table 5.1: Some Costs and Benefits for HOPCOMS and Non-HOPCOMS Users

Details	*HOPCOMS Users*	*Non-HOPCOMS Users*	*Test of Equality of Mean or Proportion*
Proportion who felt that the price they got was "fair"	92 per cent	78 per cent	–2.6***
Average transaction cost per unit of cultivated landa	Rs. 319.8	Rs. 596.1	1.55**
Average commission paid	0	9.4 per cent	–
Ratio of transactions cost to revenue from this sale	0.45	0.72	2.58***
Proportion who reported making a profit on the chosen crop this season	60.9 per cent	49.5 per cent	–1.56**
Average profit per acre earned this season	(+)Rs.1308	(+)Rs.1071	0.55
Average loss per acre for those whosuffered losses	()Rs.1525	(–)Rs. 2760	0.6
Proportion who reported being in debt this season	50.6 per cent	66.7 per cent	2.22**
Average amount of debt per acre cultivated for those who report debts	Rs.10722	Rs. 5718	1.275

* Means significant at 10 per cent; ** Significant at 5 per cent, *** Significant at 1 per cent.

Source: Kolday D, S Krishnamoorthy and S Narayanan (2008).

At present, HOPCOMS is handling about 90-100 M T of fruits and vegetables every day, Nearly 200 farmers are bringing their produces directly to HOPCOMS every day. To help the growers for transporting the produce, the Society has procurement centres in major supply areas. The procured fruits and vegetables are supplied to consumers through the Society's retail outlets in Bangalore and Kolar. Presently, HOPCOMS has 257 retail outlets (Bangalore Urban and Rural-238, Kolar-19) which provide fresh fruits and vegetables to the consumers. HOPCOMS also prepares and sells fresh juices such as Blue Grape, Mango, Orange, etc. HOPCOMS supplies inputs such as manures and fertilizers, plant protection chemicals, Gibberelic Acid, Secateurs etc. to the growers at reasonable rates (HOPCOMS, 2015).

Self Help Groups (SHGs) Based Approach

Vegetables and Fruits Promotion Council of Kerala (erstwhile Kerala Horticultural Development Programme) is the successful example of mobilizing farmers into Self Help Groups (SHGs) and federating them for better access to inputs, technology and market. The SHGs form the base units for all the interventions like Extension, Rural Credit, Group Marketing, Participatory Technology Development (PTD), Value Additions and Exports. From each SHG three master farmers are selected and trained. They lead farmers in the areas of production, credit and marketing. SHG membership enables farmers access to credit, training and to technical advice but benefits have gone beyond production related aspects (GFRAS, 2012), VFPCK follows group marketing approach where a group of 7 to 15 neighbouring SHGs will constitute a Field Centre (FC) and the SHG farmers bring their produce to a common place for marketing. Traders are coming to the Field Centers and this will increase the bargaining power of the farmers (Hall *et al*; 2004) To give additional support for bargaining, the VFPCK Market Information Centre (MIC) make available the daily market prices of banana and all other vegetables collected from different markets in Kerala and even outside. VFPCK is registered as a company where Self Help Groups of farmers account for 50 per cent of shares, Government of Kerala has 30 per cent and other related institutions hold 20 per cent of shares. A MoU was signed between VFPCK and 11 banks in the state for disbursement of credit to farmers. The SHG assess the credit worthiness and credit requirements of its members through a participatory credit planning session. At present there are around 5800 SHGs and about 111000 farmers (VFPCK, 2015).

Producer Companies

In India, producer companies are initiated based on the amendment of Company Act in 2002. In a 'Producer Company', only persons engaged in an activity connected with, or related to, primary produce can participate in the ownership. Producer companies differ from cooperatives and business enterprises in many aspects. Any ten or more individuals, each of them being a producer, that is, any person engaged in any activity connected with primary produce, any two or more producer institutions, that is, producer companies or any other institution having only producers or producer companies as its members or a combination of ten or more individuals and producer institutions, can get incorporated as a producer company (Murray, 2008).

The first producer company in India, Indian Organic Farmers Producer Company, started functioning in Kerala in 2004. Only producers with organic certification are eligible for membership of the company, where patronage for one share is fixed at Rs.40,000. An individual can purchase more than one share, but will have only one vote, irrespective of the number of shares held. The company provides advice to farmers on mapping and assessing resources (mainly soil and water), sustainable resource utilization and scientific production methods. The company markets organic products after branding. (NRAA, 2009). In Madhya Pradesh, a number of producer companies have been formed under District Poverty Initiation Programme (DPIP). Producer companies can be operational in sectors other than agriculture also. Producer companies, as they are wholly owned by primary producers and devoid of any external intervention, are supposed to be more sensitive to the priorities of small holders.

Coconut Development Board has initiated the formation of Coconut Producer Societies (CPS) by associating 40-100 coconut growers in a contiguous area with a consolidated minimum of 4000- 6000 palms (average 5000 palms). Objective of CPS is socio economic upliftment of the farmers through productivity improvement, cost reduction, efficient collective marketing and processing and product diversification. These societies are registered under Charitable Societies Act and all the societies will have Common byelaw. Establishment of societies is followed by Federations at next higher level and the Producer Companies at its apex. Through this three tier set up, Board is trying to 'inform, inspire and motivate the farmers and stakeholders on the various programmes of the central and state governments, which can be converged at farmer collectives for the betterment of coconut farmers (Anonymous, 2013).

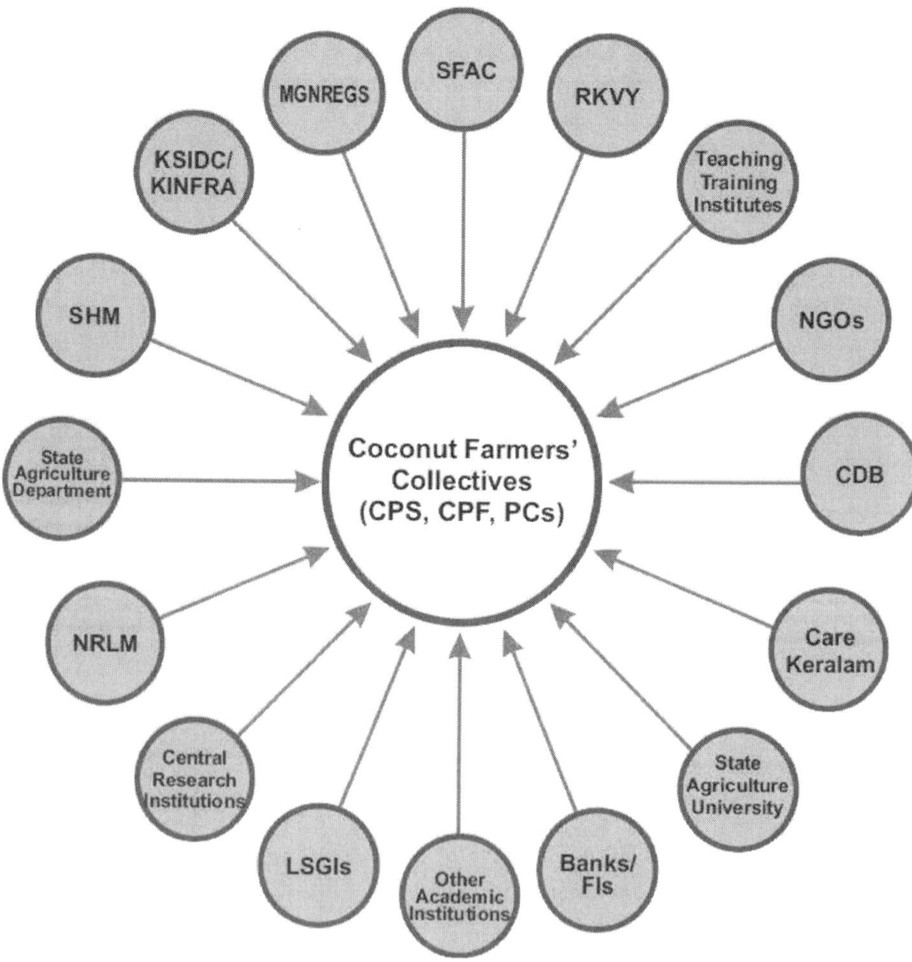

Figure 5.1: Coconut Farmers' Collective as Platform for Convergence of Various Schemes, Programmes and Agencies.

***Source*: Coconut Development Board.**

The CPSs are trying to diversify the range of product from coconut beyond the traditional copra and coconut oil nexus. The number of CPSs and Federations taking initiatives in coconut processing is increasing day by day. The most important value added product from coconut that has sought the attention of all the stakeholders in coconut is 'Neera'. The nutritious non alcoholic drink, Neera has enormous market potential. The various value added products from Neera like palm syrup; jaggery and sugar have prospects in both domestic and export market (Nair, 2013).

The tapping of Neera and its recovery is a process which requires skill and care. Skilled work force trained specifically for tapping Neera in a scientifically approved manner is required. The CPSs facilitate training of Neera technicians in coordination with the developmental programmes of the State/Central sector. Technical support for processing as well as logistic and marketing support is arranged in coordination with various agencies like Agricultural University, NHM, NRLM etc. (Anonymous, 2013).

A case analysis of Samarth Kisan Producer Company Limited, Agar in Madhya Pradesh was undertaken which strives towards enhancing income of shareholders (small and marginal farmers) by developing dynamic functional linkages with agri-business trade and industry and develop support system to enable farmers' and their institution to thrive independently in the competitive agri-business environment. It revealed that even the small and marginal famers could become seed producers. Generally a seed company makes contract with farmers having at least 2 ha land but through association with producers' company, small and marginal farmers too could draw the benefits of access to credit, supply of quality and cost-effective inputs, advisory of service providers on improved cultivation practices, and access to markets. More than 4/5th of members were highly satisfied as got technology, information, quality seed and inputs.

Implications for Extension

Considering the wide diversity in Indian agriculture, developing a blue print of extension strategy for the small holders will be a futile attempt. In an agricultural innovation system perspective, extension should look for a 'best fit' approach (Birner *et al.*, 2006) for small holders in the particular area. Such an ideal or 'best fit' approach should address the multitude of problems faced by small holders. It requires mobilization of farmers on one hand and networking with different stakeholders in agricultural innovation system *viz.* research institutions, input dealers, processors, buyers, financial agencies etc. on other hand. Public extension providers might consider repositioning themselves as facilitators of innovation networks in which their primary role is to accelerate the transmission of production information and inputs by connecting smallholders to private industry and civil society organizations. Even if technology dissemination to be effective, there has to be a set of complementary factors in place, otherwise famers will be unable to make best use of the technology and the efforts of the extension agency will be in vain. It should undertake broader roles of facilitation, information brokering, mobilization and capacity development. Reinventing extension to small holders demand operationalization of such an approach by net working with all the stake holders in the agricultural innovation system. Such an approach should be adaptive to the changes in the socio-political and economic environment. Then only it can be responsive to small holders' priorities

There is a need for new approaches to extension system as well as new skills and competencies among the extension professionals in areas of community mobilization, group formation, developing and delivering contents about market information, channelizing weather based advisory, group action, conducting intensive trainings and demonstrations on frontier technologies for quality production and value addition, enhancing linkages with service providers and agencies, educating farmers about rules and regulations about cooperatives, value-orientation and modalities for contract arrangements and arbitration, promoting awareness about risk management measures (*e.g.*, insurance schemes)etc.

References

Agarwal,Bina (2010). Rethinking Agricultural Production Collectivities. *Economic and Political Weekly*, Vol XLV No 9, February 27, 2010

Anonymous,(2013). Convergence and Collective Convergent Action. *Indian Coconut Journal*, January, 2013.

Birner, R. *et al.* (2006). From best practice to best fit: A framework for analyzing agricultural advisory services worldwide. Washington, DC: IFPRI.

Birthal,P.S and Joshi,P.K (2007). Smallholder Farmers' Access to Markets for High Value Agricultural Commodities in India, Case study #6-4 of the program: "Food policy for developing countries: the role of government in the global food system"2007, Cornell University, Ithaca, New York.

Birthal, P.S., P.K. Joshi, D. Roy and A. Thorat. (2007). Diversification in Indian agriculture towards high-value crops: The role of smallholders. *IFPRI Discussion Paper 00727*. International Food Policy Research Institute, Washington, D.C.

Chand,Ramesh, P.A. Lakshmi Prasanna, and Aruna Singh (2011) Farm Size and Productivity: Understanding the strengths of small holders and improving their livelihood. *Economics and political weekly* Vol.XLVI (25 and 26), pp 5-11.

Coulter, J., Goodland, A., Tallontire, A., and Stringfellow, R. (1999). Marrying farmer cooperation and contract farming for service provision in a liberalizing Sub-Saharan Africa, Natural Resource Perspectives, 48, ODI, London, UK.

Fischer, Elisabath and Quaim, Martin (2010) Linking Smallholders to Markets: Determinants and Impacts of Farmer Collective Action in Kenya, Discussion paper 48, Courant Research Centre, 'Poverty, Equity and Growth in Developing and Transition Countries: Statistical Methods and Empirical Analysis' Georg-August-UniversitätGöttingen.

Global Forum for Rural Advisory Services (2012). Extension Services: Case Studies, Position Paper, June 2012.

Hall, A.J., Yoganand, B., Sulaiman, R.V., Raina, R.S., Prasad, C.S., Naik, G.C.and Clark, N.G (2004). Innovations in innovation: reflections on partnerships, institutions and learning. A joint publication of the DFID Crop Post-Harvest

Programme (South Asia), ICRISAT and the National Centre for Agricultural Economics and Policy Research, India

Kaplinsky, R and Morris, M. (2000). A Handbook for Value Chain Research. IDRC, Canada.

Kolday, D., Krishnamoorthy, S and Narayanan, S. (2008) An "other" revolution? Marketing cooperatives in a new retail context: A case study of HOPCOMS in India Regoverning Markets

Kruijssen, F., Keizer M., and Giuliani, A. (2009). Collective action for small-scale producers of agricultural biodiversity products, Food Policy, 34, p.46-52.

Murray,E.V.(2008). Producer Company Model: Opportunity for Bank Finance, *CAB calling*, April-June 2008.

Nair, Deepthi,S. (2013). Unite, Think big and Win big. *Indian Coconut Journal*, January, 2013.

Narrod, Clare A.; Roy, Devesh; Okello, Julius; Avendaño Ruiz, Belem D and Rich, Karl M (2007). The role of public-private partnerships and collective action in ensuring smallholder participation in high value fruit and vegetable supply chains, CAPRi working papers 70, International Food Policy Research Institute (IFPRI).

NRAA (2009).Perspectives and Problems of Primary Producers Companies - Case study of Indian Organic Farmers Producer Company Limited, Kochi, Kerala, National Rainfed Area Authority, New Delhi, India.

NSSO. (2005). Situation Assessment Survey of Farmers – Some aspects of Farming, 59th Round (January - December 2003). National Sample Survey Organization (NSS), Ministry of Statistics and Programme Implementation. Government of India, New Delhi. Report No. 496(59/33/3).

Pandolfelli, Lauren; Meinzen-Dick,;Suseela and Dohrn, Stephan,(2007) Gender and collective action: A conceptual framework for analysis,*CAPRi working papers 64*, International Food Policy Research Institute (IFPRI).

Quizon, Antonio B. and Riguer, Mary Grace L. (2003). 'Collective Sugarcane Farming in Capiz, Philippines: Survival in a Dying Industry?'", Globalization and Small Farm Household Agricultural Practices in the Philippines. A Case Study, CARRD and ANGOC, Philippines, 2003.

Rao, V.M. (1997) Integration of production, procurement, processing, and marketing through co-operatives: a study of HOPCOMS (Karnataka). *Indian Journal of Agricultural Economics*, 1997,52(3).

Shiferaw, B., G. Obare and G. Muricho. (2008). Rural market imperfections and the role of institutions in collective action to improve markets. Natural Resources Forum 32: 25-38.

Vyas,V.S (2011) Tarlok Singh Memorial Lecture delivered at the University of Hyderabad on 14-10-2011 available at http://www.thehindu.com/todays-paper/tp-national/article2547290.ece?css=print

HOPCOMS (2015). http://www.hopcoms.kar.nic.in/; 06/02/2015

Mahagrapes (2015)http://www.mahagrapes.com/; 02/02/2015

http://mofpi.nic.in/ContentPage.aspx?CategoryId=162; 12/10/2011

NDDB(2011). http://www.nddb.org/achievement/ataglance.html; 12/10/2011

VFPCK (2015)http://www.vfpck.org/vfpck_concept.asp; 07/02/20153

Chapter 6

Boosting Rural Economies through Local Economic Development (LED): Policy Initiatives

Surya Rathore and N. Sandhya Shenoy

Division of Extension Systems Management,
National Academy of Agricultural Research Management,
Hyderabad, Telanagana

India is known as an agricultural country, as most of the population of villages depend on agriculture. Agriculture forms the backbone of the country's economy. The agricul-tural sector contributes most to the overall economic development of the country. For centuries together, the Indian village has been a self-sufficient and self-contained econ-omy. During the past forty years, rural reconstruction and development have been the major thrust of economic planning, which has caused a rapid transformation in the Indian rural economic structure.

These changes have taken place in spheres, such as land reforms, agriculture, animal husbandry, supplies and marketing, village industries, rural leadership, village administration, *etc.* With the help of the rural development pro-grammes, a cultivator is able to take advantage of the modern technological facilities in his/her agricultural operations. These cultivators are now using modern agricultural imple-ments and high-yielding varieties of seeds and fertilizers. Several other welfare services were introduced, such as opening up of schools, primary healthcare centres, improving the means of transport and communication and spread of mass media services to rural areas, *etc.* These welfare services have materially affected the rural life.

Gone are the days when Indian farmers were engaged in subsistence agriculture. In the present days, a large part of rural economy has been opened up, which has made commercialization of agricul-ture possible. The extent of commercialization of agriculture in rural economy marks the stage of its development. High production and productivity gains have turned out agriculture to be a profitable proposition. Increase in production is possible due to the use of advanced technology in agricultural operations. Massive expansion of road transport has reduced the distance between ru-ral and urban areas. Development of regulated markets and cooperative marketing structure has helped the farmers to break away from village system of moneylend-ers and middlemen. In spite of all these developments, there are a good number of people migrating from rural to urban areas in India in search of small jobs thus increasing the number of urban slums. The unfettered flow of rural migrants imposes tremendous pressure on existing infrastructures of the towns and cities creating undesirable fallout in the form of unemployment, congestion, slums and squalor, environmental degradation and spiraling crime rate. Economic plans by focusing their allocation in urban centers contributes to the problem by aggravating the urban-rural disparity which accelerates the rural to urban flow of migrants (Madhu and Uma, 2014). There are "pull" and "push" factors responsible for migration of rural people to urban areas. For example: Due to good number of employment opportunities in industries, transport, trade and other services, cities act as 'magnets' for the migrant population and attract people from outside. In other words, cities pull people from other areas. This is called pull factor. On the contrary, the "push" factor is associated with compulsion to leave villages and go to urban areas as a result of unemployment, hunger and starvation. When they do not find means of livelihood in their home villages, they are 'pushed' out to the nearby or distant towns. To promise a comfortable life to India's rural population residing in villages, it is deemed important that the push and pull factors be removed and Local Economic Development (LED) is the answer.

What is LED?

According to International Labour Organization (2012), Local Economic Development (LED) is a participatory development process that encourages partnership arrangements between the private and public stakeholders of a defined territory, typically gathered in forums that enable the joint design and implementation of a common development strategy by making use of the local resources and competitive advantage, with the final objective of stimulating economic activity and creating decent jobs for all. It involves focusing on the unique characteristics and resources of a territory and the inclusion of a variety of local actors. This process is a winning approach for self-empowered, long-term development of even remote rural areas.

Ecotourism

Ecotourism involves visiting natural areas with the objectives of learning, studying or participating in activities that do not bring negative effects to the environment; whilst protecting and empowering the local community socially and economically." Empirical surveys of tour operators and foreign tourists indicate that visitors interested in the natural beauty of the Himalayan kingdom stay longer than those who come to experience its culture. By attracting younger tourists, ecotourism could contribute to the goals of Gross Development Product. The majority of tourists and tour operators are in favour of ecotourism activities that might benefit local rural communities. Current interest in traditional food and cuisine is part of a general desire for authentic experiences. At the regional level, the dynamics of building up heritage consist in actualizing, adapting, and re-interpreting elements from the past, thus combining conservation and innovation. Local development can be seen as a process of territorial and heritage construction. Culinary heritage is a social construction and an important resource for local action. By employing local people whenever possible and paying fair wages. This prevents exploitation and benefits the local economy by preventing leakages of the revenue acquired through tourism. Many Eco tour operators are involved in working on initiatives and projects with the local community such as training and education programmes.

Agri-tourism

The agri tourism concept was formally launched by Agri Tourism Development Corporation Baramati on 1st May 2004. The concept of agri tourism envisages involvement of private sector, the farmers / Agri Tourism Service Providers based on public private partnership. The Agri Tourism Service Providers act as both hosts and guides to the visiting tourist. The Agri tourism centers have clean, hygienic environment with modern facilities for comfort of visitors. Preference is given to farm which have agricultural land attached. The Agri Tourism Service Provider is supposed to provide home cooked food, stay facilities and show the visitor the agricultural practices such as floriculture, harvesting, bee keeping, dairying *etc.* and introduce to him the village way of life through various participatory activities. The tour of village includes visiting the local artisans like the carpenter, blacksmith *etc.* The visitors have experience of festival occasion such as marriage and local *melas*. They can also participate or witness village games such as wresting, *gulidanda*, kite flying and have ride on bullock cart, tractor *etc.* Experiences such as jumping on the hay and taking bath in the tube well could be unique feature for the people and children from urban areas. They can also see the important fairs and festivals being organized in these areas along with important monuments, *havelies*, historical sites *etc.* Each farm can aim at developing a unique selling point some are specializing in organic farming, others is in floriculture, natural health management. The visitors can enjoy the natural surroundings in fresh air. The visitor gets an exposure of local community life which means attending a panchayat meeting, exposure of local songs, food, dances, art and craft *etc.* Community participation is a very important part of this concept as the scheme envisaged visit to the village, interaction with panchayat members and attending village functions *etc.* As our experience has shown the tourists cannot be looked after by the Farmers alone. The spillover effect is witnessed by whole of the village community. In most of the group packages, numbers of activities are undertaken such as bullock ride, artisans display, *mehandi*, folk music and dance, village safari, gift article shop *etc.* This has resulted in community participation of the villages, generation of employment and additional income for the villagers.

There are yet number of challenges associated with this profession in India.A broad range of policies and regulations affect the operations and viability of most farms operations. The main policy and regulation challenges facing the agri-tourism industry are as follows:

1. There is a lack of understanding at all government levels concerning how existing regulations and policies impact agri-tourism operations.
2. No existence of policies and regulations with respect to the appropriate size of facilities and the use of land for agri-tourism activities.
3. Policies are perceived to be developed without consultation with farm operators and without an analysis of the potential impacts on agri-tourism operations.
4. There is a lack of consistency in regulation interpretation by government administrators.
5. There is limited awareness and understanding of existing (if either) agri-tourism policies and regulations amongst farmers. As well, many farmers are unaware of how to obtain information on these regulations and how to apply them to their specific operations.

Agripreneurship

Agriculture remains a key sector of the Indian economy accounting for 13.2 per cent share in the gross domestic product (GDP) and about 13 per cent of the total export earnings. Since the inception of New Economic Reforms, adoption of liberalization, privatization and globalization (LPG) and world trade organization (WTO) in 1992 – 95, it is expected that rural area will grow at par with urban area. A shift from agriculture to agribusiness is an essential pathway to revitalize Indian agriculture and to make more attractive and profitable venture. Agripreneurship have the potential to contribute to a range of social and economic development such as employment generation, income generation, poverty reduction and improvements in nutrition, health and overall food security in the national economy. Agripreneurship has potential to generate growth, diversifying income, providing widespread employment and entrepreneurial opportunities in rural areas.

Self Help Groups

The existing formal financial institutions have failed to provide finances to landless, marginalized and disadvantaged groups. The decade of 1990's witnessed growth of various people's organizations. In this context the role of Self-help groups (SHG's), especially of women has assumed a critical challenge. Self-help groups approach is the key element of social mobilization. Linked with micro-finance, the SHG approach and movement has now been accepted as an effective intervention strategy for poverty alleviation. The origin of SHG's could be treated to mutual aid in Indian village community. SHG is a viable organizational setup to disburse micro credit to the rural women for the purpose of making them entrepreneur and encouraging them to enter into entrepreneurial activities. Credit needs of the rural women can be fulfilled wholesomely through the SHG's. The women led SHG's have successfully demonstrated how to mobilize and manage thrift, appraise credit needs, maintain linkages with the banks and enforce financial self-discipline. SHG's enhance the equality of status of women as participants, decision-makers and beneficiaries in the democratic, economic and social and cultural spheres of life. They encourage women to take active part in the socio-economic progress of the society.

Traditional Rural Craftsmanship

Traditional crafts, according to UNESCO, are products made completely by hand, or with the help of simple tools, and reflect the creativity, the culture and the heritage of the craftsperson. Although often viewed as a small scale industry, traditional crafts comprise world market of US $235Bn and, according to UNESCO; handicrafts represent 20 per cent of the rural activities of developing countries, contributing to at least 3 per cent of their GDP. While the major markets for traditional crafts are the developed countries, the developing countries contribute to the largest share of the low end market segment of the crafts. Over the past few years, rapid industrialization has lured many craftsmen and artisans away from practicing their core skills towards higher rewarding jobs. This has resulted in the loss of the cultural patronage and expertise from an industry that relies heavily on skills and craftsmen. The craftsmen have not benefited greatly from the move to industrial jobs due to their lack of expertise in that domain. Though, there are some challenges associated with this trade but due to its increasing demands from international markets, this rural business would promote cultural heritage and creativity of local artisans, and, in the process helping local communities in their socio-economic development.

Organic Farming

Organic farming is a sustainable production, which highlights the harmonic connection of humans and their environment; it harmonizes the developed agricultural methods with the facilities of nature. It combines the ancient practical knowledge of the conventional farming with today is scientific results. The looking after of our natural habitat and organic farming are related and they affect each other since according to researches organic production in nature protected area or near to it increases the efficiency of both of them. On the other hand organic farming is nothing but applied environmental protection and since the environmental state of the Hungarian agriculture is better than that of the more developed countries in Western-Europe, it gives a comparative advantage to the country if it uses environment friendly methods due to their higher efficiency. The National Agri-Environmental Protection Programme was launched to help organic production and environmental protection: to protect natural sources, product quality and to support food safety. Organic farming can be connected to any function of the rural area: Its economic role is represented by widening the range of employment, creating new workplaces, improving the population keeping ability of the area. ï Its ecological function is well connectable to environment- and landscape conservation and hence in an indirect way to maintaining biodiversity. Social and cultural functions mean in this context the revitalizing and developing of traditional farming methods. Organic (eco-) farming delivers an ever growing market background for producers in rural areas. It is important; however that organic farming has to be concerned as a complex and integrated system. It means on one hand that the shift to organic farming should include not only the production of organic raw materials but also the processing, packaging and marketing of them. On the other hand it is important to establish the organizational and interest forms and systems which cover the production itself, the equipment and genetic base supply, small and medium size processing facilities, packaging, advisory and marketing management. Regarding growing domestic and foreign consumption trends, good price positions, lack of export restrictions organic farming can turn itself into a new Ñalternativeî employment segment in many areas of the country. In this way the previously ignored factor of rural development can potentially turn into a dynamiting engine of regional development. The total annual production of organic food in India in the last financial year was 120,000 tons. The Indian states involved in organic farming in India are as follows:

- Gujarat
- Kerala
- Karnataka
- Uttarakhand
- Sikkim
- Rajasthan
- Maharastra
- Tamil Nadu
- Madhya Pradesh
- Himachal Pradesh

The main organic agricultural products are bajra, mustard, wheat, chilly, pulses, maize, ginger, soybean, large cardamom, passion fruit, Dungarpur pulses, mustard, til, Nagour guar-cumin, guar-wheat, moong, mustard, Ganganagar cotton, Jaisalmer, bajra, Bharatpurbajra and wheat *etc.*

Para-vets (Pashu Mitra)

Para veterinary workers are those people who assist a veterinary physician in the performance of their duties, or carry out animal health procedures autonomously as part of a veterinary care system. The job role varies throughout the world, and common titles include veterinary nurse, veterinary technician, veterinary assistant and veterinary technologist, and variants with the prefix of 'animal health'. Para-vets are community based animal health workers. served to fill a large gap in extension services and have enabled more people to access vital information and services to protect their livestock. Para vets are able to charge for their services and this is also a valuable source of income for local people.

Policy Initiatives to Encourage LED

- To encourage the Stakeholders to promote and practice Ecotourism practices, the Ministry of Tourism has included categories of awards " Best Eco friendly Hotel", 'Best Responsible Tourism Project", "Best Eco friendly

Practices by Tour Operators" in the National Tourism Awards presented annually to various segments of travel, tourism and hospitality sector

- ✰ The Indian Agri-tourism Council (**IAC**) should initiate programmes which increase awareness and understanding of the Agri Tourism amongst governments and farmers.
- ✰ Teach traditional crafts in schools as vocational education. We believe this is a crucial step in ensuring the knowledge spill-over to future generations. Considerable work has been done in some countries, particularly at an introductory level. This will generate an interest that will result in awareness and also provide the required local patrons of the projects. However, it is important to extend the school education to a technique-oriented vocational education. One of the challenges is to have teachers who are very well versed in handicrafts, and are able to extend their passion and interest to the coming generations. Also important is that the training and education gives rise to subsequent employment in the industry.
- ✰ Development of partnerships between agri-tourism and value-added processing businesses and other types of businesses should be encouraged at the regional level. Develop networking partnerships between researchers and agri-tourism organizations to provide the farmers with information on product and marketing partnerships occurring worldwide. The Internet is a tool which can facilitate this networking of ideas and information.
- ✰ Encouraging medium and small scale industries in rural India by means of waiving taxes. This would help the rural youths to get job in their village itself thus, promoting LED.

References

Bairwa, S. L; Lakra, K; Kushwaha, S; Meena, L. K. and Kumar (2014) 'Agripreneurship Development as a Tool to Upliftment of Agriculture' Internatinal Journal of Scientific and Research Publications, Vol. 4 (3), pp. 1-4.

Boosting local economies in rural areas, Rural Policy Briefs, International Labour Office 2012

Gurung, D B and Seeland, K (2008) Ecotourism in Bhutan: Extending its benefit to Rural Communities Annals of Tourism Research Vol. 35 (2), pp. 489-508.

http://www.ecotourdirectory.com/ecotourism (accessed on 15/2/2015)

http://www.tourism.gov.in (accessed on 15/2/2014)

Madhur, G. R. and Uma, H. R. (2014) Rural to Urban Migration – Opportunities and Challenges, International Journal of Advanced Research Vol. 2 (6), pp. 389-394.

Sarudi, C; Szakaly, Z; Mathe, A and Szento, V (2003) The Role of Organic Agriculture in Rural Development, Agric. Conspec.Sci. Vol. 68 (3), pp. 197-202.

Chapter 7

Information and Communication Technology for Dissemination of Agricultural Knowledge through Krishi Vigyan Kendras in India

M.S. Meena[1], S.K. Singh[2], Shobhana Gupta[3] and R.B. Kale[4]

[1]*Principal Scientist (Agricultural Extension)*
[2]*Director;* [4] *Scientist (Agricultural Extension)*
ICAR-Agricultural Technology Application Research Institute, Zone-VI, Jodhpur, Rajasthan
[3]*Deputy Director Extension, RVSKVV, Gwalior, M.P.*

Introduction

Agriculture is the most important sector of Indian economy. It contributes 13.9 per cent of India's Gross Domestic Product (GDP). Nearly 59 per cent population derives its livelihoods from the agricultural sector. In spite of an alarming increase in the population, it has managed to attain self sufficiency in food grain production. However, looking into the total area under cultivation, achievements made in agricultural productivity are far below the desired levels. One of the major constraints is the deficiency in transfer of relevant knowledge from National Agriculture Research System (NARS) and State Agriculture Departments to small and marginal farmers who form bulk of the nation's agricultural population (Patil *et al.*, 2011). In India, there are about 120 million farm holdings and the number is growing year by year. There is need to provide one village extension personnel for 800-1000 farm families whereas, the requirement of field level extension worker is estimated to be about 13-

15 lakh, against the present availability of only one lakh extension worker (Planning Commission, GOI, 2007). Estimates indicated that 60 per cent of farmers do not access any source of information for advanced agricultural technologies resulting in huge adoption gap (NSSO, 2005). In this scenario, inadequate technical manpower is to be compensated by the extensive use of Information and Communication Technologies (ICTs).Land and water resources are almost reaching at their limits; hence, achieving food security heavily relies on knowledge resource. In this scenario, ICT can complement the traditional extension system for knowledge resource delivery to the millions of the farmers. ICTs are required to facilitate stronger linkages with research-extension-client system. The feedback received through ICTs to be more accurate and faster. The ICT tools such as internet and mobile networks have the potential to provide agri-information services that are affordable, relevant (timely and customized), up-to-date, high accessibility and farmer friendly. To fasten the agricultural extension activities, different organizations have been started many initiatives in India.

ICT Initiatives for Agricultural Development in India

ICTs in agriculture have the potential to facilitate greater access to information that drive or support knowledge sharing. Some initiatives in India that use ICTs for agricultural development are:

Table 7.1: Some ICT Initiatives for Agricultural Development in India

Sl.No.	*Organizations*	*Projects Undertaken*
1.	**Central government sponsored**	
	Department of Agriculture and Co-operation's National Agriculture Technology Project.	
	AGMARKNET	
	Community Information Centres	
	Warana Wired Village Project of Maharashtra: Application of satellite communication for training to field extension workers in rural areas (Indian Space Research Organization)	
2.	**State government sponsored**	
	Rajasthan	Janmithra
	Madhya Pradesh	Gyandoot
	Andhra Pradesh	e-Seva (Computer-Aided Online Registration Department),
	Karnataka	Bhoomi (Land Record Computerisation), Raitha Samparka Kendra Online, Marketing and CAD in Northern Karnataka
	Tamil Nadu	Rasi, Miyams Karchipular
	Gujarat	Seva (Automated Milk Collection Centres of Amul dairy cooperatives)
	Maharashtra	Setu, Online Complaint Management System - Mumbai
	Kerala	E-Srinkala
3.	**Corporate sector sponsored**	
	ITC-IBD	E-Choupal
	EID-Parry(Nagarjuna group of companies, AP)	ikissan Kendra, Tata Chemicals Chirag Kendra

Contd...

Table 7.1–Contd...

Sl.No.	*Organizations*	*Projects Undertaken*
4.	**Non-government organizations and other private sector organizations**	
	Information Village Research Project in Pondicherry	
	Agriwatch.com	
	Drishti.com	
	Tarahaat.com by Development Alternatives	
	Mahitiz-samuha	
	VOICES by Madhyam Communications	

Source: Yogisha, Ramappa, Mahajanshetti, and Vijayakumar, 2008.

Krishi Vigyan Kendra (KVKs) is an important pillar of public agriculture extension system in India. The main functions of KVK are transfer of technologies from research lab to farmers' field, capacity building of farmers through organizing trainings and extension activities and serve as knowledge resource center for the farmers. It is the only institutional system at the district level for technological backstopping in agriculture and allied sectors. Many initiatives have been taken to integrate ICTs in KVKs to carry out extension activities in timely and effective manner. Main initiatives are:

Kisan Mobile Advisory Services

Information technology offers immense potential in enhancing the efficiency and effectiveness of agriculture extension programmes and dissemination of the best practices. However, it is to be taken care that we have very low internet penetration and as a matter of fact the rural internet penetration is restricted to only a few states and regions. The mobile based farmer advisory system is the best alternative solution. For instance, with the help of ICT tools, Kisan Mobile Advisory Service (KMAS) has been introduced in Zone-VI covering 20 districts in Rajasthan and 5 districts in Gujarat. Now, it is expanded in the each KVK of Zone-VI. Similarly each KVK in Zone VII has enrolled approx. 5000 farmer/extension officers under KMA for sending SMS for overcoming major issues in time with discipline based information to farmers. Facility has also been provided for receiving the problems of farmers and informing the solution of common and specific problems. The message incorporates weather alerts, market information and technical advisory and consultancy. The market and weather information is sourced from Internet, interpreted and broadcast according to the varying needs of the farming community.

Virtual Krishi Vigyan Kendra (vKVK)

Virtual Krishi Vigyan Kendra (vKVK) is a platform that connects KVKs with farmers through internet and mobile technology. The objective here is to bridge the gap between the farmers and the KVK expert.vKVK was developed by IIT-Kanpur in a consortium mode as part of ICAR-NAIP supported project entitled "Engaging Farmers, Enriching Knowledge-Agropedia Phase II". vKVK service started in August,

2011 as a pilot project in four states with the official partners viz International Crops Research Institute for the Semi-Arid Tropics (ICRISAT)-Hyderabad, University of Agricultural Sciences-Dharwad and ZPD-IV. Initially, the service was started in 24 KVKs and was up-scaled in 190 KVKs covering more than 30,000 farmers. The vKVK service is cutting across all the service providers and can be accessed flawlessly on even low end mobile handsets. All vKVK services are free of cost to the farmers.

vKVK is a service created specially to help the KVK scientist and farmers to communicate through cell phone like devices. The scientist can send SMS and voice messages to the registered farmers. These messages are classified according to crops and agro climatic zones. The farmers get targeted alert and advisory messages based on the registration information they have provided to their KVK. KVK provides voice-messaging service that acts as automatic voice dialer that sends the recorded voice messages in the form of automatic phone calls to the farmers. This service is provided to all farmers irrespective of telecom network. Presently this service is being made available only to the registered farmers at free of cost.

Information Hub through V-sat Connectivity (ERNET)

The Indian Council of Agricultural Research has provided e-linkage to selected KVKs and Zonal Project Directorates during XI plan. These KVKs have been connected electronically through VSAT and connected to the KVK hub established at Agricultural Extension Division, New Delhi. The communication between KVKs across the country was enabled through IP phones provided to each KVK. Using this facility, e-seminars are webcasted from the KVK hub involving experts on various topics of interest to KVKs.KVK Pathanamthitta is one among these KVKs in the country with information hub with V-sat connectivity established by ICAR. The facilities at hub are used by farmers, students and extension functionaries to a great extent. KVK also organize orientation training programme on information technology to its farmers and other stake holders along with training programmes held at KVK. This enables users to familiarize with the ICT and its application in agriculture. The presentations telecasted regularly from KVK hub team at Delhi play a vital role in updating not only the expertise at KVK but also many extension functionaries and farmers.

Almost All Questions Answered (aAQUA)

aAQUA is the IIT Bombay product implemented with KVK Baramati as knowledge partner. Almost All Questions Answered is a Farmer Knowledge Exchange available at www.aaqua.org answering questions from progressive farmers in 4 languages in any one of 420 districts in India and some places abroad. Any farmer, agriculturist or hobbyist can register and post questions and a panel of agriculture experts answers questions based on the problem description and photos, if any. Contextual information such as geographical location, weather, and season are retrieved automatically and made available to experts. Currently, questions may be asked in one of four languages - Hindi, Marathi, Kannada and English. Originally developed in the Developmental Informatics Lab of IIT, Bombay, aqua uses relational database management systems and information retrieval techniques with query optimization, intermittent synchronization and multilingual support. aAQUA Mini the mobile version of aqua can help farmers get information to phone upload photos and get expert advice and SMS alerts directly to the farmers. Mobile phones are now affordable and there seems little doubt that cellular telephones will cover a large percentage of rural population in the country over the next few years.

Community Radio

Radio as an affordable and sustainable ICT technology will not only supplement the information dissemination processes of KVK, but will also offer a platform of 'accessible' media to the farmer community who can share their experiences with a wider audience on all agriculture related issues. Community participation, volunteerism and part management of the station by the community are the contributing factors for the success of CR. Community radio (CR) is an attempt to make best use of the most widely available but least explored communication technology, the radio. KVKs offer radio broadcast services for farming communities, encourage farmers to listen and to provide their feedback on the agriculture broadcast. This helps KVKs to improve their services further based on the needs of the farmer communities CR reaches out to various people simultaneously and in diverse locations such as farms, while fishing out in the sea, at local shops, public places, as long as radio is within the reception range. The end user just needs to have an FM radio to listen to the broadcast. Farmers and other representatives of the community can also participate in program production. CR can, thus, play an immense role as an ICT for development (ICT4D), especially in rural India where other ICTs have limitations. However, policy changes and guidelines in favour of KVK are very important the success of CR projects, as KVK on its own has very limited powers and must constantly work within the guidelines stipulated by ICAR. Examples of CRs operated by KVKs such as KVK in Sangaria in Rajasthan, KVK Junagadh in Gujarat, KVK Baramati, KVK Babhaleshwar, in Maharashtra. The types of programmes telecasts on CR includes Drama, Features,Interviews, Panels and discussions, Documentary, Editorial, Vox Populi (Voice of people) etc.

Agropedia

Agropedia is an ICAR initiative which is a semantically enabled agricultural knowledge hub. Agropedia being an ICT solution for Indian agriculture, has state-of-art enabled platform to aid both certified knowledge holders (scientists, researchers, teachers, students, extension workers) as well as Tacit knowledge holders (progressive farmers, businessman) to contribute, share and search agricultural content. agropedia is a read/write web, where the users of agropedia can not only read its content but also can add new posts, edit the existing wiki page, comment on any existing page and also can rate the content. agropedia main components are Knowledge models, Content, Social Networking. agropedia is not only restricted to web, it also extends itself to many offline deployment solutions like mobile, appliance, hand held devices which means agropedia can be accessed even at places which has no or low connectivity. Knowledge Models are the basis of agropedia. This is the part responsible for the semantic enablement for the content. Knowledge Models are used to catalogue the content which is further used for indexing and hence power retrieval of the data.

This is built using cmap tool. Content in agropedia in mainly classified in two categories: 'GYANDHARA' and 'JANAGYAN'.

Gyandhara is a sub repository of all certified content. This section is called as Extension Material (http://agropedia.iitk.ac.in/?q=content/extension-material) in agropedia, which is further divided into Library, Crop Calendars, Dos and Don'ts and Package of Practices type of content. 'Library' is a collection all certified knowledge which is uploaded by various agricultural universities. 'Crop Calendars', 'Dos' and Don'ts' and 'Package of Practices' are where information like for crop variety which practice is to be followed in which season. For the other category of content Janagyan, as the name itself suggest, it is a form of emergent knowledge. This sub repository of agropedia is socially co-created content through one these sub section: agrowiki, agroblog, Forum and agrochat. agropedia. Agropedia is being accessed by more than 150 countries with 350 unique visits per day having 3300 registered user across globe and possess more than 7000 pages in English, Hindi Kannada, Telugu and many other languages.

Openagri is an open access agricultural research repository. This is a research focused space on top of agropedia. openagri is a state-of-art website which allows many content types to be added like Journals, Books, Book Chapters, Conference Proceedings, Conference Papers, Multimedia, Theses, Reports, Abstract, Preprints and Miscellaneous. The content when uploaded here goes to the repository and a software called Agrotagger, runs in the background taking the uploaded file as an input and it then automatically generates keywords from a controlled vocabulary called Agrotags.

Agrotags is a vocabulary developed by IITK and ICRISAT with collaboration of FAO, Rome. It is actually a subset of the agrovoc which is a huge agricultural thesaurus containing several thousand terms in many international languages. It has some 40.000 terms in English language alone. Out of these given terms, Agrotags consists of approx 3,054 terms which are handpicked terms out the 40,000 agrovoc terms. They are selected based on their usage at tagging model. Agrotagger is software which accepts a document and automatically generates Agrotags/agrovoc/keywords for the given document. It is currently deployed on the openagri platform and also available for generic usage at: http://agropedialabs.iitk.ac.in/Tagger/

New Extension Tool for Technology Transfer: Social Media

Agricultural extension and advisory services currently face several new and complex challenges. Social Media can play an important role in enhancing interactions and information flows among different actors involved in agricultural innovation and also enhance capacities of agricultural extension and advisory service providers. These are highlighted below drawing examples of its use in extension and advisory services which are primarily led by voluntary efforts of few individuals, groups and professional networks. Social media is fast-becoming an integral part of the marketing landscape, and the relatively low cost and wide reach – when used effectively – can be a godsend for cash-strapped techecnology transfer offices that need to enhance their promotional efforts without many budget dollars to work with. Blogging, Twitter, Facebook, LinkedIn, YouTube and many other social media avenues offer near limitless opportunities to push your available innovations in front of a captive audience. Just like any other traditional marketing channel, the keys to success are in the planning and execution. That's why *Technology Transfer Tactics* has brought together an international panel of experts who are successfully utilizing social media to garner attention for their technologies, create an active community surrounding their innovations, and bring more licensees in the door.

Social media refers to the means of interactions among people in which they create, share, consume and exchange information and ideas in virtual communities and networks. Kaplan and Haenlein (2009) define social media as "a group of Internet-based applications that build on the ideological and technological foundations of Web 2.0 and that allow the creation and exchange of user-generated contents". Social media has changed the way we communicate, read, search, think, talk, watch, listen, and sometimes start a revolution—be it political and or social. Social media is more about sociology and psychology of communication than about technology. If Facebook was a country, it would be the second most populated in the world after China. The fact that only young people and teenagers are most active in social media is turning into an urban myth. Irrespective of age, gender, religion, geography, people are entering in the world of social media. The major social Medias like. Facebook, YouTube, Google+, LinkedIn, Twitter, Whatsaap, Blogs etc are gaining popularity worldwide.

Facebook

With over one billion profiles globally, Facebook enables its exceptionally active user base to freely share information about their lives, particularly through pictures and video. Users access a wide array of functions, including phone, chat, payment, and login to meaningfully connect both through Facebook's website and across the internet via plugins on other websites. Facebook provides a number of useful features for community building. First, it is an extremely helpful tool for meeting new people, especially 'friends of friends' who may be interested in connecting by association. Communities can be grown via Pages, Events, and Groups. Though fluid, Pages are typically used to showcase specific 'brands,' while the functionality of Groups better supports discussion and community. Events are a useful and easy way to gain attention for your community happenings.

YouTube

YouTube is a video-sharing website. The service was created by three former PayPal employees in February 2005. In November 2006, it was bought by Google. YouTube now operates as one of Google's subsidiaries. The site allows users to upload, view, and share videos, and it makes use of Adobe Flash Video and HTML5 technology to display a wide variety of user-generated and media video. Available content includes video clips, TV clips, music videos, and other content such as video blogging, short original videos, and educational videos.Most of the content on YouTube has been uploaded by individuals, but media corporations and other organizations offer some of their material via YouTube, as part of the YouTube partnership program. Unregistered users can watch videos, and registered users can upload videos to their channels.

Table 7.2: Use of Facebook in Agriculture Extension (on 7th December, 2014)

Name of Group/Community/Pages	*Description*	*Target Users*	*Region*	*Followers/ Members*
	Use by farmers			
Livestock Information and Marketing Centre (https://www.facebook.com/groups/Livestock.TN/	Members (farmers, extension personnel, scientists, market functionaries, consumers, local leaders, *etc.*) of this group share information related to livestock production, management, marketing, *etc.* A separate page is also on facebook related only to marketing of livestock. (https://www.facebook.com/Livestock.Market)	Agricultural stake-holders related to livestock	Tamil Nadu, India	49 483
Mkulima Young (Young Farmer) (https://www.facebook.com/mkulima.young)	This page is an information sharing platform for young farmers started Joseph Macharia, a young farmer himself. Mostly agro-advisory and market information are shared.	Young farmers	Kenya	39 082
Natural farming Development Centre (https://www.facebook.com/groups/NaturalFarmingTN/).	Members of the group share information related to organic farming, permaculture, hydroponics, aquaponics, Natural Repellents, *etc.*	Farmers interested in organic and zero budget agriculture	Tamil Nadu, India	16 268
Turmeric Farmers' Association of India (https://www.facebook.com/turmeric.farmers)	This page was created by turmeric farmers to stabilize price of turmeric in the market. Till date, the farmers connect through the page and share information to keep turmeric price stable and increase marketing opportunities of turmeric.	Turmeric farmers	India	2 911
National Ecological Producers Association (APNE) (https://www.facebook.com/anpe.peru)	Information related to ecological farming is shared through the page.	Farmers	Peru	3 061
Use by extension centres				
Krishi Vigyan Kendra, Namakkal (https://www.facebook.com/krishi.namakkal)	Krishi Vigyan Kendra, Namakkal communicates information related to farmers' training programmes, availability of inputs *etc.*	Subject Matter Specialists of KVK, farmers, and other agricultural stakeholders	Namakkal, Tamil Nadu, India	1464

Contd...

Table 7.2–Contd...

Name of Group/Community/Pages	*Description*	*Target Users*	*Region*	*Followers/ Members*
	Use by extension professional networks			
Agricultural Extension in South Asia (AESA) (https://www.facebook.com/groups/428431183848161/)	Members post links to relevant publications on extension and advisory services, announcements of workshops and conferences, major policy decisions on extension, reports of meetings and workshops relevant to the broader theme of extension	Agricultural Extension stakeholders	South Asia	7 550
Global Forum for Rural Advisory Services (GFRAS) https://www.facebook.com/groups/gfras/	This page provides information related to advocacy and leadership on pluralistic, demand-driven rural advisory services.	RAS Professionals and others	Global	1 794
	Use by extension personnel			
Vivasayam Karkkalam	Mr. Madhu Balan, a public extension officer started facebook group to cater the	Farmers and others those	Tamil Nadu,	12 118

Google+ (or Google Plus)

Google+ (or Google Plus) is a social networking and identity service that is owned and operated by Google Inc. Google has described Google+ as a "social layer" that enhances many of its online properties, and that it is not simply a social networking website, but also an authorship tool that associates web-content directly with its owner/author. In October 2013, Google counted 540 million active users who used at least one Google+ service, of which 300 million users it deemed active. A 2013 survey found that 30 per cent of surveyed smart phone users used the Google Plus app at least once a month. Some examples are #agriculture-Google+, New Holland Agriculture-Google+, UNFAO-Google+ etc.

LinkedIn

LinkedIn is a business-oriented social networking service. Founded in December 2002 and launched on May 5, 2003,it is mainly used for professional networking. The basic functionality of LinkedIn allows users (workers and employers) to create profiles and "connections" to each other in an online social network which may represent real-world professional relationships. Users can invite anyone (whether a site user or not) to become a connection. LinkedIn has 225 million global members and is a tremendous resource for job seekers and head hunters. The reason to list LinkedIn here is that it has cultivated a collection of 1.3 million Groups, but most social activity congregates in just 5 percent of them. The best Groups are very active places for conversation and drive a disproportionate amount of referral traffic to outside sources. LinkedIn is an oft-overlooked resource to cross-post on, build project reputations, and boost traffic to other platforms. Examples of use of LinkedIn in agriculture are:

- ✩ Rural Development Network
- ✩ Young Professionals' Platform for Agricultural Research for Development (YPARD)

Twitter

At the beginning of 2013, it was widely reported that Twitter was the fastest growing social network in the world, and is available in 25 languages. The service enables users to send short 140-character messages ("tweets") to each other along with linked and images. Roughly a quarter of those messages are re-tweets, where one user shares another's message with their own followers. Tweets are generally public, with non-registrants being able to read them on the web-based platform, unless user private preferences 'lock down' an account for approved followers only. Twitter is the go-to place for breaking news (and unconfirmed rumors) around the world, accessible by searching specific event or topic-specific hash tags, which 'trend' as they grow in popularity. In the agriculture and development communities, Twitter has taken on a special role as a 'town square' of sorts; where connected users need not be bi-directional 'friends,' but can instead opt to 'follow' each other in a one-way fashion. Examples of use of twitter in agriculture are:

- ✰ @agriGoI is the twitter account used by the Department of Agriculture and Cooperation, Ministry of Agriculture, Government of India.
- ✰ @WFP, World Food Programme
- ✰ @FAO statistics,
- ✰ @IFFCO_PR

WhatsApp

WhatsApp Messenger is a cross-platform instant mobile messaging app which allows you to exchange messages without having to pay for SMS. WhatsApp Messenger is available for iPhone, BlackBerry, Android and Windows Phone because WhatsApp Messenger uses the same internet data plan that is used for email and web browsing so there is no cost to message sent or received. In addition to basic messaging, WhatsApp users can create groups, send each other unlimited images, video and audio media messages.

ResearchGate

ResearchGate is a social networking site for scientists and researchers to share papers, ask and answer questions, and find collaborators, access scientific knowledge, and make their research visible. It is a network dedicated to science and research.

Blog

A blog (a truncation of the expression weblog) is a discussion or informational site published on the World Wide Web and consisting of discrete entries ("posts") typically displayed in reverse chronological order (the most recent post appears first). Until 2009 blogs were usually the work of a single individual, occasionally of a small group, and often covered a single subject. More recently "multi-author blogs" (MABs) have developed, with posts written by large numbers of authors and professionally edited According to critics and others Blogger is the most popular blogging service used today, however Blogger does not offer public statistics.

Websites

A website, also written as web site or simply site is a set of related web pages typically served from a single web domain. A website is hosted on at least one web server, accessible via a network such as the Internet or a private local area network through an Internet address known as a uniform resource locator (URL). All publicly accessible websites collectively constitute the World Wide Web. The pages of a website can usually be accessed from a simple Uniform Resource Locator (URL) called the web address. The URLs of the pages organize them into a hierarchy, although hyper linking between them conveys the reader's perceived site structure and guides the reader's navigation of the site which generally includes a home page with most of the links to the site's web content, and a supplementary about, contact and link page.Some websites require a subscription to access some or all of their content. Many KVKs have started their websites to provide the information to their clientele *i.e.* farmers as well as extension officials. Some examples of KVK websites are www.kvkcard.org; www.kvkdatia.com; www.kvkujjainzpdvii.org etc.

Future Outlook

Despite the huge potential of harnessing ICT for agricultural development, only a few isolated projects have been initiated in India and in other parts of the world. Interestingly, many of these projects were initiated by NGOs, private organizations, cooperative bodies and government organizations rather than by government-established agricultural departments. This shows the apathy of agricultural development departments towards incorporating ICT in their day-to-day activities. To formulate a strategy for overall agricultural development, isolated ICT projects need to be studied and the experience gained must be documented in order to draw lessons for the future. On the other hand, the need to market agricultural produce at competitive prices will also change the farmers' attitude towards usage of ICT. ICT will thus help to sustain Indian agriculture. It is necessary to develop ICT based agricultural services along with a communication backbone (such as a fiber optic network) in rural areas. Though the use of ICT in agriculture is in a nascent phase in India, ICT has immense potential to standardize and regulate agricultural processes and address the needs of farmers. It will therefore definitely serve as an important tool for agricultural development in the near future.

References

1. Diem, K.G., Hino, J., Martin, D. and Meisenbach, T. (2011). Is extension ready to adopt technology for Journal of Extension (6). delivering programs and reaching new audiences?, Article number FEA1,49www.joe.org/joe/2011december/a1.php.

2. Kaplan and Haenlein. (2009). Harnessing social media in agriculture. Report for New Zealand Nuffie Scholarship Fund.

3. Lin, Thomas (16 January 2012). "Cracking open the scientific process". *The New York Times*. Retrieved 26 June 2014.

4. Meena, M.S.; Gupta,S. Rohilla, P.P. and Kale, R.B. (2015). Harnessing ICTs for Dissemination of Agricultural Knowledge through KVKs: Way Forward. Lead Paper presented in ISEE National Seminar on Extension Innovations and Methodologies foe Market-led Agricultural Growth and Development, during Feb 26-28, 2015 at RVSKVV, Gwalior.

5. Newbury, Elizabeth, Humphreys, Lee, and Fuess, Lucas (2014) Over the Hurdles: Barriers to Social Media Use Journal of Extension, in Extension Offices, Article number 5FEA152(5).www.joe.org/joe/2014october/a1.php

6. NSSO, (2005). Access to Modern Technology for Farming, Situation Assessment Survey of Farmers. 59th Round Report No. 499, National Sample Survey Organization (NSSO), Ministry of Statistics and Program Implementation, Government of India, New Delhi.

7. Patil, V.C., Hugar, L.B., Priya, P., Prabhuraj, A., Balagi, V. and Yaduraju, N.T., 2011, Information and Communication Technologies for Agriculture Knowledge Management in India. World Applied Sciences Journal, 14(5):794–802.

8. Planning Commission, GOI (2007) Recommendations of Working Group on Agricultural Extension for Formulation of Eleventh Five Year Plan (2007-12). Planning Commission, Government of Indiahttp://www.fao.org/docs/eims/upload//257364/Patil_presentation.pdf

9. Yogisha, G.M., Ramappa, K.B., Mahajanshetti, S.B., and Vijayakumar, H.S. (2008). Information and Communication Technologies (ICTs) in India: An insight into progress of rural economy. A National Journal of Agriculture and Rural Development, 40(1), 34-36.

Chapter 8

Indigenous Practices Followed by the Farmers for Storage of Food Grains

M.C. Ahire, S.H. Ban, D.N. Pharate and A.B. Gaikwad

Department of Extension Education,
MPKV - College of Agriculture, Dhule – 424 004, Maharashtra

Farmers store food grains by using different storage structure and use various practices in household. While adopting these practices, the farmers face many constraints. Based on this fact to find out best indigenous practices followed by the farmers for food grain storage and constraints faced by them in it.The present study was conducted in Akole tahsil of Ahmednagar district (M.S.). The data collected from 120 respondents. Regarding knowledge of farmers about the storage of food grains it is revealed that majority of the respondents were from medium knowledge level category. All the farmers had knowledge of prevention of food grain from moisture, curative measures for control of stored grain pest, control measures for stored grain pest of pulses. Regarding indigenous practices followed by farmers for storage of food grains it is revealed that majority of the farmers stored paddy seed in gunny bags (Bhople, R.S. and Darbha, 2002), all the farmers using mud and cow dung for keeping food grain air tight and majority of the farmers used neem leaves for protecting food grain from pest. Most of the farmers had not adopted the practices namely use of fumigants, spraying of insecticides on storage materials before storage of food grains. Majority of the farmers had fully adopted the practices namely sun drying and cleaning of harvested food grain. The important constraints reported by considerable number of farmers is that lack of knowledge regarding identification of stored grain pest, unavailability of different storage structure, high cost of storage structure, inadequate place for storage (Chavan K.M, 2009), skill persons for making storage structure are not available for constructing storage structure. The study has indicated farmers possessed least knowledge about the practices such as name of fumigants, control of rice moth and so on, the extension agencies should educate the farmers by providing knowledge and skills though demonstrations and trainings regarding these storage practices and needs to motivate the farmers for the successful adoption of these practices in order to avoid losses from storage. The findings of the study also confirmed that majority of the farmer had fully followed the practices, which were known to them, which were simple and inexpensive and which could be accommodated within the resources available with them. The practices which were not known to them and which were expensive, beyond their resource limits were not followed by them.

Keywords: *Indigenous practices, Storage, Food grains, Constraints.*

Introduction

Food grains comprising of cereals, millets and pulses, form the primary and staple food of majority of the population in India. For fetching higher prices to the food grains in the market not only production of grains is important but storage of grain is very important. It is reported that about 30 per cent losses in food grains is due to unavailability of storage facilities, and lack of knowledge regarding storage. The production of food grains is seasonal, but their consumption is uniformly spread throughout year; hence the agricultural commodities need to be stored.

The sustained increase in production brought within the problems of storage. Millions of tonnes of valuable grains worth more than 2000-3000 crores of rupees are either damaged or lost for want of knowledge of scientific methods of storage, Every year due to various agencies like insects, rats, moisture and birds losses are seen. The loss is not merely in terms of quantity but also in quality of food grains. The qualitative loss is attributed to chemical changes in protein, carbohydrates, amino acids, fatty acids and vitamins which affect the nutritive value of the grains. Protection of food supplies through sound scientific storage practices is thus, a need to pay urgent attention. There are various agencies working in agriculture, who are undertaking storage functions *viz.* farmers and government organization *etc*. Farmers themselves undertake the storage at farm level, for their own consumption. At farm level different type of storage structures are observed depending upon commodity, duration of storage, climate, cost of storage structure, incidence of rodents, insects etc.

Proper storage of the food grains in rural families then would not only help in bridging the existing marginal food gap, but would also contribute to health and nutrition by conserving grain quality. Most farming communities have traditional methods and structures for storing their produce. These vary based on the climatic conditions, type of crop and construction materials available. Indigenous methods of storage also include use of plants with insect repellent properties, sun drying, hanging the produce in bundles from roof top etc. There are different ways like Indoor, Outdoor and Underground storage practices. Farmers store food grains by using different storage structure and use various practices in household. Various practices are followed by farmers for storage of food grains while adopting these practices of food grain storage, the farmer faces many constraints. Based on this fact to find out best indigenous practices followed by the farmers for food grain storage and also to document these practices and to identify the constraints of these farmers the study has undertaken with following specific objectives.

1. To study the indigenous knowledge of farmers about the storage of food grains.
2. To study the indigenous practices followed by farmers for storage of food grains.
3. To study the constraints faced by the farmers in storage of food grains.

Methodology

The study was conducted in Akole tahsil of Ahmednagar district of Maharashtra. From the Akole Tahasil, 10 villages were selected randomly. From every selected village, 12 farmers were selected. Thus in all from 10 villages 120 respondent were selected for study. Data was collected with the help of structured schedule by personally interviewing the farmers. Frequency, percentages were worked out for analysing and interpretation of data.

Results and Discussion

1. Knowledge of the Farmers about Storage of Food Grains

The data from Table 8.1 revealed that all the farmers had Knowledge of Prevention of food grain from moisture (100 per cent), Curative measures for control of stored grain pest (100per cent), Control measures for stored grain pest of pulses (100per cent), followed by Preventive control measures for stored grain pest (97.50 per cent), Keeping of storage structure air tight(97.50 per cent), Method to keep storage structure air tight(97.50 per cent), Nature of damage by rice weevil (97.50 percent), followed by Two names of the stored grain pest (95.83 percent), Proper drying of food grains(95.00 per cent), Reasons behind burring of dead rats in the soil (95.00percent), Name of rodenticides (75.00 percent), Preparation of poison bait for rat(71.66 per cent), Keeping of rat poison bait at proper place (71.66 per cent), Mechanical control of rat (66.66 per cent).However, it was observed that only few respondents had knowledge about name of insecticides (1.66 per cent),percentage of the moisture in food grain at the time of storage(6.66 per cent), precautions at the time of use of fumigants (9.16per cent).

2. Different Containers Used by the Farmer for Storage of Food Grains

The data from Table 8.2 revealed that majority (70.83 percent)of the farmers have reported that they stored paddy in *gunnybags*, followed by *Kothali* (66.66 percent), *Tattaya Bamboo* (35.00 percent), *Tank*(12.50 percent), *Kangi*(10.83 percent), *Balad*(6.66 percent), *Tattaya Iron* (4.16 percent), *Earthen pot* (1.66 percent). For storage of wheat (22.50 percent) respondents used *gunny bags, Tattaya* (Bamboo) (5.00 percent), *Tank*(4.16 percent), *Kothali* (2.50 percent),*Tattaya* (*Iron*) and *balad* (1.66 percent). For storage of bajra (41.66 percent) respondents used *gunny bags, Kothali* (1.66 percent), *Tattaya*

Table 8.1: Knowledge of the Farmers About Food Grain Storage Technology (N=120)

Sl.No.	Particulars	Number Yes	Per cent No
1.	Two names of the stored grain pest	115 (95.83)	5 (4.16)
2.	Preventive control measures for stored grain pest	117 (97.50)	3 (2.50)
3.	Percentage of the moisture in food grain atthe time of storage	8 (6.66)	112 (93.33)
4.	Prevention of food grain from moisture	120 (100)	–
5.	Proper distance between floor and storage bags	29 (24.16)	91 (75.83)
6.	Keeping of storage structure air tight	117 (97.50)	3 (2.50)
7.	Method to keep storage structure air tight	117 (97.50)	3 (2.50)
8.	Name of insecticides	2 (1.66)	118 (98.33)
9.	Curative measures for control of stored grain pest	120 (100)	–
10.	Nature of damage caused by rice weevil	117 (97.50)	3 (2.50)
11.	Control of rice moth	15 (12.50)	105 (87.50)
12.	Control measures for stored grain pest of pulses	120 (100)	
13.	Proper drying of food grains	114 (95.00)	6 (5.00)
14.	Precautions at the time of use of fumigants	11 (9.16)	109 (90.83)
15	Nature of damage by rat	36 (30.00)	84 (70.00)
16.	Mechanical Control of rat	80 (66.66)	40 (33.33)
17.	Name of the rodenticides	90 (75.00)	30 (25.00)
18.	Preparation of poison bait for rat	86 (71.66)	34 (28.33)
19.	Keeping of rat poison bait at proper place	86 (71.66)	34 (28.33)
20	Precautions at the time of use of rodenticides	102 (85.00)	18 (15.00)
21	Reasons behind burring of dead rats in thesoil	114 (95.00)	6 (5.00)
22.	Proper Temperature at the time of storage of food grain	65 (54.16)	55 (45.83)
23.	Proper Humidity at the time of storage of foodgrain	56 (46.66)	64 (53.33)
24.	Information regarding improved methods forstorage of food grains	70 (58.33)	50 (41.66)

Table 8.2: Different Containers Used by the Farmer for Storage of Food Grain (N=120)

Sl.No.	Particulars of Container	No.	Capacity (Quintal)	Paddy	Bajra	Wheat	Pulses	Other
A.	**Food grains**							
1.	Gunny bags	104	1	85 (70.83)	50 (41.60)	27 (22.50)	14 (11.66)	25 (20.83)
2.	Kothali	96	2-3	80 (66.66)	2 (1.66)	3 (2.50)	–	–
3.	Tattaya (Bamboo)	54	10-15	42 (35.00)	6 (5.00)	6 (5.00)	–	–
4.	Tank	53	1-1.5	15 (12.50)	–	5 (4.16)	33 (27.50)	–
5.	Iron bin	47	0.5-1	–	–	–	45 (37.50)	2 (1.66)
6.	Kangi	13	1-3	13 (10.83)	–	–	–	–
7.	Balad	10	10-15	8 (6.66)	–	2 (1.66)	–	–
8.	Tattaya (Iron)	7	10-15	5 (4.16)	–	2 (1.66)	–	–
9.	Earthen pot	5	1-2	2 (1.66)	–	–	3 (2.50)	–
B.	**Seed**							
1.	Gunny bags	121	1	75 (62.50)	5 (4.16)	18 (15.00)	9 (7.50)	23 (19.16)
2.	Iron bin	95	0.5-1	–	–	–	66 (55.00)	–
3.	Kangi	21	1-2	21 (17.50)	–	–	–	–

Figure in parentheses indicate percentage.

Bamboo(5 percent). Maximum number of farmer used *iron bin* (37.50 percent),*Tank* (27.50 percent), *Gunny bags* (11.66 percent), *Earthen pot* (2.50 percent) for storage of pulses. For other grains most farmers used *gunny bags* (20.83 percent). For paddy seed storage (62.50 percent) of the respondents used *gunny bags* followed by *Kangi* (17.50 percent).While pulse seed was stored in *iron bin*(55.00 percent) followed by *gunny bags* (7.50 percent).

This means that respondents had used locally available material for storage.These findings are similar to the findings of Anonymous (2000), Bhople and Darbha (2002), Chinnaiyam *et al.* (2001), Uplap (2003), Chavan (2009).

Table 8.3: Traditional Methods Used for Storage of Food Grains

Sl.No.	*Practices followed*	*Respondents (N= 120)*	
		Number	*Percentage*
A.	**Methods to keep food grain air tight**		
1.	Use of mud and cow dung	106	83.33
2.	Iron lid	66	55.00
3.	Plastic cover	54	45.00
4	Bamboo lid	15	12.50
B.	**Prevention of food grain from moisture**		
1.	Use of stand	109	90.83
2.	Keeping distance between bags and ceiling roof	75	62.50
C.	Material used for preparation of storage structure		
1.	Mud and cow dung	110	91.66
2.	Leaves	98	81.66
3.	Bamboo	75	62.50
4.	Plastic paper	22	18.36
D.	**Material used in storage of food grains**		
1.	Neem leaves	116	96.66
2.	Ash	108	90.00
3.	Salt	44	36.66
4.	Oil	–	–
5.	Boric powder	4	03.33
E.	**Care of store places**		
1.	Storage place should be clean	109	90.83
2.	Plugging the hole, crack, crevices in storage room	99	82.50
3.	Keeping proper distance between two rows of gunny bags	81	67.50
4.	Prevention of gunny bags from wall	40	33.33
5.	Storing food grains at proper height	33	27.50
6.	Proper height of gunny bags	24	20.00

3. Traditional Methods Used for Storage of Food Grains

A. Methods to Keep Food Grain Air Tight

The data from Table 8.3 revealed that all the farmers were using mud and cow dung (83.33 per cent), Iron lid (55.00per cent), Plastic covering (45.00 per cent) followed by Bamboo lid (12.50 per cent).

B. Prevention of Food Grain from Moisture

The data showed that farmer followed a practices namely Use of stand (90.83 per cent), Keeping distance between bags and ceiling/roof (62.50 per cent).

C. Material Used for Preparation of Storage Structure

The data indicated that for preparation of storage structure materials like use of mud and cow dung (91.66 Percent), Bamboo (62.50 per cent), Leaves (81.66 per cent), Plastic paper (18.33 per cent).

D. Material Used in Storage of Food Grains

The data indicated that large majority (96.66 per cent) of the farmer were using neem leaves for protecting food grain from pest. However, (90.00 per cent) respondents were using ash, (36.66 per cent) respondents were using salt, and relatively small proportion (3.33 per cent) respondents were using boric powder.

E. Care of Store Places

The data indicated that majority (90.83 per cent) of respondents keeping storage place should be clean, 82.50 percent plugging the hole, crack, crevices in storage room, 67.50 percent keeping proper distance between two rows of gunny bags, followed by few of them Storing food grains at proper height(27.50 per cent), Proper height of gunny bags (20.00 per cent).These findings are similar to the findings of Anonymous (2000), Bhople and Darbha (2002),Uplap (2003), Natarajan and Santha (2004), Chavan (2009).

4. Adoption of Food Grain Storage Practices by Farmer

A. Preventive Measures

The data from Table 8.4 showed that majority of the respondent completely adopted the practice of sun drying (96.66per cent), proper drying and cleaning of harvested food grain on threshing yard (93.33 per cent). It was also observed that (56.66per cent) of the respondents 'partially' adopted the practice of keeping gunny bags away from ceiling, (52.50 per cent) keeping gunny bags away from wall, followed by (50.83 per cent) sieve the food grains after threshing, Use of stand (50.00 per cent), Cleaning of storage place (50.00 per cent), Keeping Proper aeration in storage room (50.00 per cent).

B. Curative Measures

The data from Table 8.4 showed that only few respondents 'Completely' adopted the practice spraying of insecticides on storage materials before storage of food grains (2.50 per cent), 5.83 per cent farmer partially adopted this practice. However (5.83 per cent) of the respondents completely adopted the practice of use of fumigants EDB/ EDCT/Aluminum phosphide, (4.16 per cent) respondents 'partially' used this practice,

Table 8.4: Adoption of Food Grain Storage Practices by Farmer

Sl.No.	Practices	Adoption		
		No	Partial	Complete
1.	Proper drying and cleaning of harvested food grain on threshing yard	–	8 (6.66)	112 (93.33)
2.	Sieving the food grains after threshing	5 (4.16)	61 (50.83)	54 (45.00)
3.	Heat			
	1) Sun drying	–	4 (3.33)	116 (96.66)
	2) Keeping proper moisture content during storage of food grains	33 (27.50)	58 (48.33)	29 (24.16)
4.	Prevention from moisture			
	1) Use of stand	22 (18.33)	60 (50.00)	38 (31.66)
	2) Keeping gunny bags away from ceiling	22 (18.33)	68 (56.66)	30 (25.00)
	3) Keeping gunny bags away from wall	23 (19.16)	63 (52.50)	34 (28.33)
	4) Cleaning of storage place	10 (8.33)	60 (50.00)	50 (41.66)
	5) Keeping proper air space in storage room	15 (12.50)	60 (50.00)	45 (37.50)
B.	Curative measures			
1.	Spraying of insecticides on storage materials before storage of food grains	110 (91.66)	7 (5.83)	3 (2.50)
2.	Use of fumigants EDB/EDCT/Aluminium phosphide	108 (90.00)	5 (4.16)	7 (5.83)
	Sun drying	50 (41.66)	25 (20.83)	45 (37.50)
C.	Maintenance of storage place			
1.	Proper aeration	11 (9.16)	37 (30.83)	72 (60.00)
2.	Not storing old and new food grains together	17 (14.16)	53 (44.16)	50 (41.66)
3.	Keeping gunny bags in rectangular shape in the storage room	50 (41.66)	40 (33.33)	30 (25.00)
4.	Plugging hole crack, crevices in the storage room	27 (22.50)	45 (37.50)	48 (40.00)
D.	Control from rodents			
1.	Rearing of cat	58 (48.33)	–	62 (51.66)
2.	Use of trap	26 (21.66)	60 (50.00)	34 (28.33)
3.	Use of rodenticides (Zinc Phosphide/Aluminium phosphide	25 (20.83)	53 (44.16)	32 (26.63)

(37.50 per cent) 'Completely' adopted repetitive sun drying of grains to avoid nature of damage due to pest, (20.83 per cent) 'partially' used this practice. It was observed that majority of the respondents not adopted the practices namely use of fumigants (90.00 per cent), spraying of insecticides (91.66 per cent) on storage materials before storage of food grains.

C. Maintenance of Storage Place

The data from Table 8.4 showed that majority (60.00 per cent) of the respondents were 'Completely' practicing the method of keeping proper aeration in the storage room, (30.83 per cent) partially adopted this practice. However, (41.66 per cent) of the farmer fully practicing the method not storing old and new food grains together, (44.16 per cent) partially adopted this practice. Only (25.00 per cent) fully adopting the method keeping gunny bags in rectangular shape in the storage room, (40.00 per cent) respondents were 'Completely' adopting plugging hole crack, crevices in the storage room, (37.50 per cent) partially adopted this practice.

D. Control from Rodents

The data from Table 8.4 indicated that majority (51.66 per cent) of the farmer had reared cat and (28.33 per cent) of them fully followed the practice of using trap, (50.00 per cent) partially using trap. However, 26.63 per cent of the respondents fully used the practice use of rodenticides (Zinc Phosphide/Aluminum phosphide), (44.16 per cent) partially adopted this practice. It is clear that use of insecticides and fumigants are not adopted by large majority of farmer. Thus extension agencies and development workers in the region should plan and implement suitable strategy to increase the adoption of these storage practices.

5. Constraints Faced by the Farmer for Storage of Food Grains

A. Lack of Knowledge

The information in Table 8.5 revealed that majority (82.50 per cent) of the farmer faced constraints about Lack of knowledge regarding identification of stored grain pest followed by chemicals and their proportion in use (58.33 per cent), lack of knowledge for storage of food grain (45.83 per cent), lack of information in storage (41.66 per cent), nature of damage by stored grain pest (36.66 per cent), nature of damage due to diseases (35.00 per cent).

B. Unavailability of Material

Table 8.5 revealed that majority (59.16 per cent) of the farmer faced constraints like unavailability of different storage structure followed by unavailability of chemicals for storage (57.50 per cent) and unavailability of spray pump and dusters (18.33 per cent).

C. Economic

Table 8.5 revealed that majority (90.00 per cent) of the farmer faced constraints in respect of high cost of storage structure followed by unavailability of loan (40.83 per cent), high cost of chemical (34.16 per cent).

Table 8.5: Constraints Faced by the Farmers for Storage of Food Grains

Sl.No.	Practices	Respondents	
		Number	Percentage
A.	**Lack of knowledge about**		
1.	Identification of stored grain pest	99	82.50
2.	Chemicals and their proportion in use	70	58.33
3.	Lack of knowledge for storage of food grains	55	45.83
4.	Lack of information in storage	50	41.66
5.	Nature of damage by stored grain pest	44	36.66
6.	Nature of damage due to diseases	42	35.00
B.	Unavailability of		
1.	Different storage structure	71	59.16
2.	Unavailability of loan	69	57.50
3.	High cost of Chemicals	22	18.33
C.	**Economic**		
1.	High cost of storage structure	108	90.00
2.	Unavailability of loan	49	40.83
3.	High cost of chemical	41	34.16
D.	**Place**		
1.	Inadequate place for storage	49	40.83
E.	**Other**		
1.	The skill persons for making storage structure are not available for constructing storage structure	41	34.16
2.	Bamboos are not easily available	22	18.33

D. Constraints Regarding Place

Table 8.5 revealed that farmer faced constraints of inadequate place for storage (40.83 per cent) in food grain storage practices.

E. Other

Table 8.5 revealed that farmer faced constraints in the skill persons for making storage structure is not available for constructing storage structure (34.16 per cent), Bamboos are not easily available (18.33 per cent).The study has brought out some important constraints in storage of food grains. The concerned agencies should take suitable steps for reduce the intensity of these constraints in the region.

These findings are similar to the findings of, Uplap (2003) and Chavan (2009).

Conclusions

The study has indicated the knowledge level of the farmers regarding food grain storing practices. The farmer possessed least knowledge about the practices such as name of insecticides, precautions to be taken at the time of use of fumigants, control of rice moth and so on, The extension agencies should educate the farmer by providing knowledge and skills though demonstrations and trainings regarding these storage practices and needs to motivate the farmers for the successful adoption of these practices in order to avoid losses from storage. The findings of the study also confirmed that majority of the farmer had fully followed the practices, which were known to them, which were simple and inexpensive and which could be accommodated within the resources available with them. The practices which were not known to them and which were beyond their resource limits were not followed by them. These findings suggested that there is need to drive farmer successfully through the adoption process of stored grain practices. The extension organizations may do this by using appropriate extension teaching methods at the respective steps of adoption process. Lack of knowledge about identification of stored grain pest was the major knowledge constraint faced by the farmer. It is, therefore, suggested that the extension agencies may take suitable measures to provide the technical know-how of this pests to the farmer. Unavailability of storage structures and high cost of storage structures, were the other major constraints faced by the farmer in storage of food grain. It is, therefore, suggested that improved storage structures at subsidized rates.

References

Anonymous, 2000. Documentation of indigenous technical knowledge. Unpublished report. Konkan KrishiVidyapeeth., Dapoli.

Alam, M., Islam, M.O. and Mirza Hasamuzzaman. 2009, Performance of Alternate Storage Devices on Seed Quality of Boro Rice. Middle- East Journal of Scientific Research 4(2): 78-83

Bhople, R.S. and Sudha Darba. 2002. Indigenous Grain Storage Practices Followed in Rural Household. Agricultural Extension Review 12: 26-29

Borkar, M.M. and Rasekar, A.K. 1999. Adoption of Grain Storage Practices by Farm Families. Maharashtra J. Extn. Edu. 18: 252- 254

Chavan, K.M. 2009. Food Grain Storage Practices Followed bythe Farm Women. M.Sc. (Agri.) Thesis (unpublished)submitted to Dr. Balasaheb Sawant Konkan KrishiVidyapeeth., Dapoli.

Chinnaiyan, P. Selvaraj, K.N. and Srinivas, N. 2001. A study onfarmers knowledge on seed storage management inTamilnadu. Indian. J. of marketing: 8-12.

Natarajan, M and Santha Govind. 2004. Indigenous Post Harvest Management Technologies. LEISA INDIA 6 no 3: 12.

Uplap, P.J. 2003. Adoption of Food Grain Storage Practices by Farm Women and their Training Needs. M.Sc. (Agri.)Thesis (Unpublished) submitted to MPKV., Rahuri.

Chapter 9

Realizing Farmer's Perception: Utmost Important in Adoption Process of Fish Farming Technology

Biswarup Saha

Department of Fisheries Extension, College of Fisheries, Central Agricultural University (CAU), Lembucherra, Tripura

This study was conducted to investigate how realizing farmer's perception has an influence on the adoption process of fish farming technology in Tripura. An *ex-post facto* research design in the purposively selected districts *i.e.*; South Tripura and West Tripura were employed to collect data from 200 respondents randomly sampled from 10 villages from those districts. Instruments used for data collection included semi structured interview schedule, Participatory Rural Appraisal, researchers' observations and secondary information sources. The overall adoption index in the study area was found 59.74. Farmers who realized their objective intensified the activity and were able to produce 412 kg/kani (1kani=2.5 Acre/0.16ha) production that was considered more profitable than other crop production. Conversely, those who did not realize their objectives abandoned or practiced low level technology. This finding suggests that farmer's objective and priority should be known before introduction of the technology and efforts should be directed towards realizing the objectives. Additionally, efforts should be made to improve production technology, preparation and preservation skills and harvest strategies.

Keywords: *Fish farmers, Perception, Technology, Adoption gap.*

Introduction

Tripura is a landlocked hilly state in northeastern India with altitudes varying from 15 to 940 m above sea level. It is the second smallest state in area of around 10,492 sq. km. (after Sikkim), but second most populous state (after Assam), among the eight land-locked north-eastern states, is categorized as resource poor state with heavy pressure on land and low normal productivity because of its soil being mostly acidic. Fisheries is an important economic activity in Tripura, both from an economic and a nutritional point of view, providing income, employment and food security to the people. Around 95 per cent of population in Tripura love to consume fish. The state is holding first rank in per capita fish consumption as inland state in the country. Moreover, Tripura reported to be highest in number of households (per thousands) consuming fish and in connection with price it has been shown that the state experienced highest retail price of fish per kg. in India. Around 1.35 lakh people depend on fisheries for their livelihood. The existing total aquatic resources of the state estimated to be 24,704 ha during 2012-13. (Fisheries Statistics 2012-13, GoT). However, current level of fish produced in the state is not able to meet the burgeoning demand and hence large amount of fish is being brought from the mainland state i.e; West Bengal, Andhra Pradesh, Bihar and even from the country Bangladesh. The total production of fish in the state during 2012-13 was about 58,655 metric tonnes and imported around 23,660 metric tones from. Per capita availability of fish was around 16.34 kg as against the requirement of 22.92kg. The scope for horizontal expansion in aquaculture is limited in the state due to hilly topography as well as wide coverage of forest area. Therefore, to reduce the gap between demand and supply and to achieve fish self-sufficiency in the state, Importance is given to explore the vertical expansion in fish production.

According to FAO (1996) the main objective for introducing fish farming in the study area was to: increase cash income and/or animal protein component in the diet of rural communities, achieved through small-scale aquaculture integrated with suitable farming system. Another objective was a pond being an efficient means of using surplus resources, thus increasing efficiency of farm activities (ibid.). The objective of improving food and income is considered more decisive to the adoption process of fish farming technology. While the rationale for adopting fish farming as stipulated above is clear, few if any researchers have attempted to assess how adopter's objectives have been realized. Most studies have concentrated on production technology giving a lesser priority to the extent to which the technology has attained farmer's objectives for undertaking it. Lack of balanced assessment of the two has contributed to poor understanding of why the adoption rate and level of intensification are low, and the abandonment rate is high. Review of different studies suggests that adequate attention has not been given to the study on how realizing farmer's objective has influence on the adoption process of fish farming technology. The objective of this paper therefore is to make a thorough investigation on how realizing farmer's objective and perception has influenced the adoption, abandonment and intensification of fish farming technology in the study area.

Methodology

The study was conducted out using ex-post facto research design in the purposively selected districts *i.e.;* South Tripura and West Tripura. Two blocks from each district were selected purposively and five villages were selected from each block using by simple random sampling method. Ten fish farmers were selected from each village using simple random sampling method. Thus, a total 200 fish farmer constituted the respondents for the study. The instruments used for data collection were a structured interview schedule, Participatory Rural Appraisal (PRA) and Researchers observation conducted in each village with a group of farmers. The interview schedule was prepared to solicit information on objectives for undertaking fish farming and assessed whether the objectives were realized or not. In addition, secondary information was collected from various sources and was used to design questionnaires and explaining a big part of this study.

Conceptual Framework to Realizing the Farmers' Adoption Process

For the purpose of this study it is assumed that farmers make adoption decision based upon utility consideration (Batz *et al.*, 1999). When a technology is introduced in a given area, the decision whether to adopt, continue and intensify it or not will depend, among other things, on how a technology meets household objective of food and income security. Household resources (*i.e.* land, labor, cash income, on-farm inputs and knowledge/skill) are allocated across various activities based on their contribution to household food and income security (Figure 9.1). The allocation of resources is often characterized as a two-stage process in which the first priority is given to meet food security requirement (Temu, 1999). The second objective is then to maximize income using the remaining resources (ibid.). Fish farming contributes to household needs through provision of fish for home consumption and cash income. However, fish farming is not the only activity, which meets the above objectives; there are other activities, which compete with fish farming to meet those objectives. For instance, it competes with livestock production to meet animal protein intake and competes with crop production, livestock production and off-farm activities to earn cash income (Figure 9.1). That means fish farming competes with those activities for resources to realize household objectives. The decision whether to adopt, continue or intensify fish farming technology or not will depend on how best and easy the technology will accomplish the intended objectives.

Result and discussion

The age of the respondents ranged from 19 to 60 years with mean age of about 42.40 years. The results presented in Table 9.1 revealed that most of the respondents (88.5 per cent) were in their late adulthood. The literacy level of the respondent is very high, *i.e.*, 96.5 per cent. It was observed from the Table 9.1 that more than half of the sampled respondents (51.5 per cent) had low to lower medium level of yield (up to 205 kg/kani), followed by around 48 per cent respondents obtained more than 205 kg/kani yield in their pond. The average fish yield in the study area is around 285.7 kg/kani (1kani=2.5 Acre/0.16ha) as compared to the scientific composite fish culture

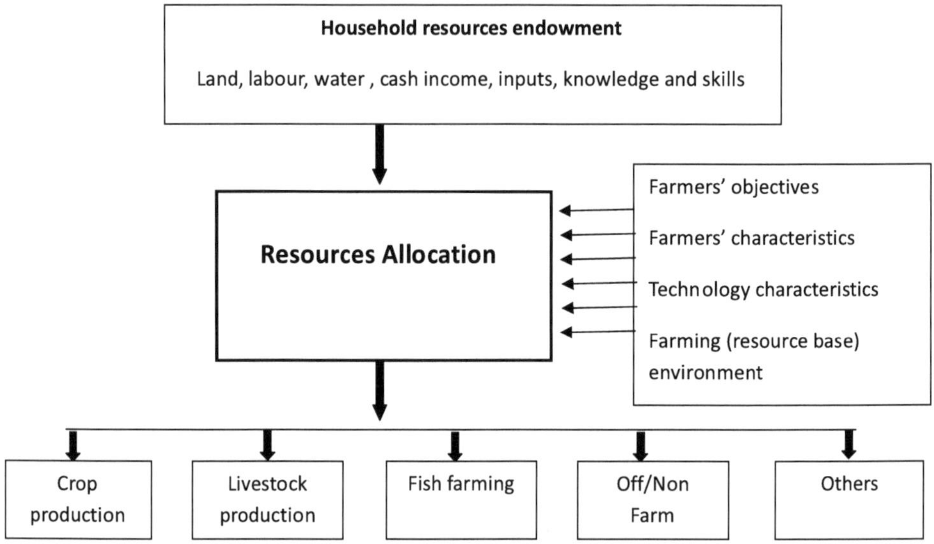

Figure 9.1: Farmers Decision to Adopt, Continue or Intensify Fish Farming Technology.
***Source*: Wetengere (2008).**

(*i.e.*, 400 kg per kani) (Anonymous, 2002). The possible reason for this trend might be due to the fact that the sampled farmers did not realize the objective of fish farming and thus did not adopt the improved practices of fish production. More than 75 per

Table 9.1: Demographic Characteristics of the Respondents

Respondents Characteristics		*Frequency*	*Percentage*
Age	Late adolescence (18-25 years)	6	3.0
	Early adulthood (26-40 years)	17	8.5
	Late adulthood more than 40 years	177	88.5
Experience in aquaculture	Least (upto 5 years)	24	12
	Medium (5-10 years)	30	15
	High (more than 10 years)	146	73
Yield	Low (upto 205.33 kg/kani)	50	25
	Lower medium (205.34-265 kg/kani)	53	26.5
	Higher medium (265.01-316kg/kani)	49	24.5
	High(>316.01kg/kani)	48	24
Pond size	0.01-1 acre	151	75.5
	1.01-2 acre	28	14
	2.01-3 acre	11	5.5
	3.01-4 acre	4	2
	4.01-5 acre	2	1
	5.01 acre and above	4	2

cent of the respondents having pond size of 0.01-1 acre followed by 14 per cent having 1.01-2 acre. The average size of the surveyed water bodies was found 0.73 Acre.

Farmer's Objectives for Undertaking Fish Farming:

The Figure 9.2 shows that more than two third of the sampled households adopted fish farming in order to obtain fish for home consumption and to generate cash income. This result is similar to those by FAO (1996), Edwards *et al.* (1997), Wetengere *et al.* (1998) and Wetengere (2000). This implies that household resource allocation is often characterized as a 2 – stage process in which first priority is given to meet food security requirement. The second objective is then to maximize income using the remaining resources. This is consistent with finding by Temu (1999).

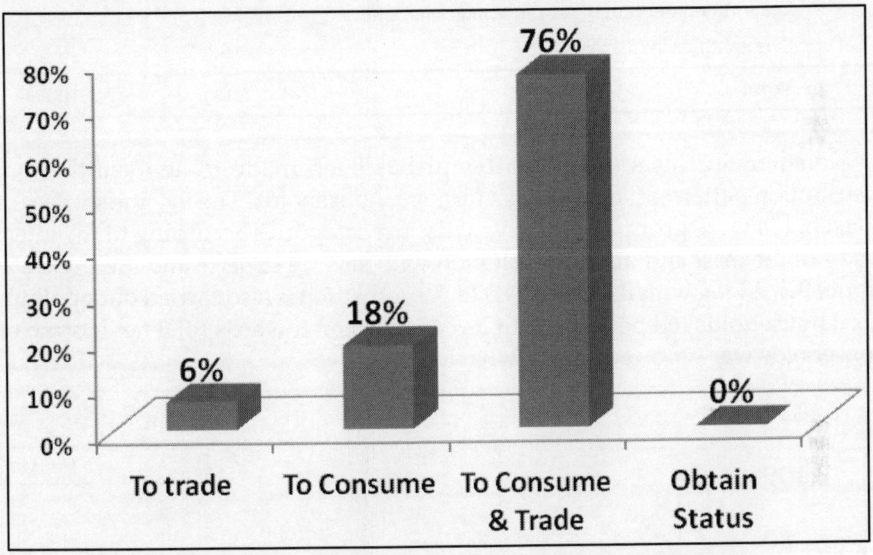

Figure 9.2: Farmer's Objectives for Undertaking Fish Farming.

Realization of the Objective of Fish for Home Consumption/Food Security

It is observed from the study that all the selected farmers are fish eater. Participants in a Participatory Rural Appraisal (PRA) conducted in this study were of the opinion that although most farmers adopted fish farming technology for both food and cash generation; it was the former objective which was more pressing and important in the study area. Farmers' perception of adoption of fish farming for home consumption was ranked first by 80 per cent respondents, ranked second by 14 per cent and the rest by 6 per cent. The importance small-scale farmers attach to household food security shows that food production is a number one priority.

Realization of the Objective of Generating Cash Income/Income Security

It is clearly visible from the Table 9.2 that majority of the fish farmers (54 per cent) earned less than 50 per cent of their earning from fish farming. It has been found that many farmers did not take fisheries in serious earning point of view.

Table 9.2: Distribution of Sampled Households as per Share of Fisheries to Total Income

Sl.No.	*Share of Fisheries to Total Income*	*Frequency*	*Percentage*
1.	More than half	92	46
2.	Between one half and one quarter	63	31.5
3.	Less than one quarter but more than one tenth	29	14.5
4.	One tenth or less	16	8
	Total	200	100

Further, in this study, an attempt has been made to analyze the income distribution pattern of sampled fish farmers' households. The income sources of the households were grouped into seven distinct categories according to the intensive study of the area and after consultation with several experts and are presented in Figure 9.3. Along with fisheries, livestock production is also carried out by almost all rural households (68 per cent), but its contribution towards total net income of the households was reportedly somewhat lower.

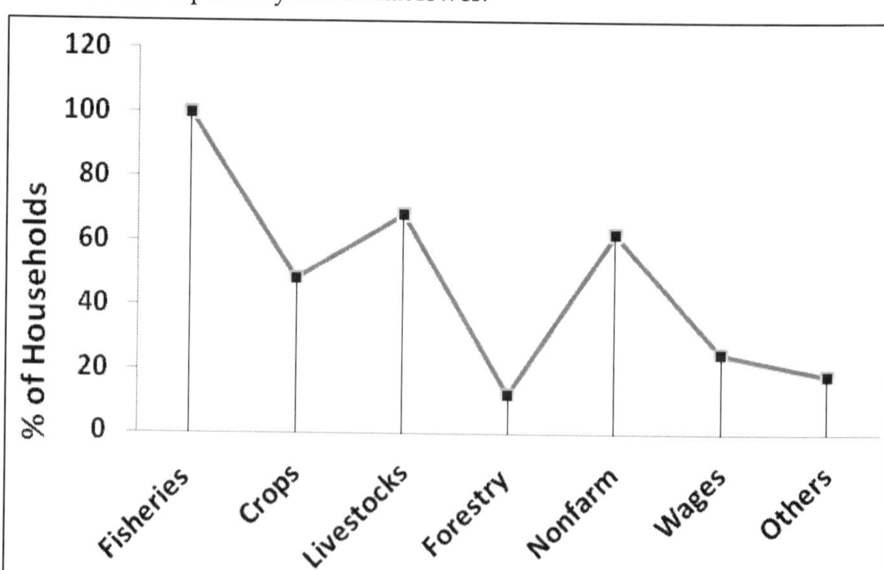

Figure 9.3: Different Sources of Income among Farmers' Households.

The main reason behind that the households generally kept unproductive local breeds. Their businesses out look from the livestock sources were also very lower.

Around 48 per cent of the fish farmers practiced crop production. The major crops were reportedly rice and vegetables. More than 60 per cent of the households participated in different non-farm income activities. In this connection, Saha (2006) reported that poor farmers generally adopt the non-farm income as their secondary or tertiary income sources as a part of a 'coping' strategy. Of the seven cash income earning activities in the study area, fish farming ranked sixth in terms of frequency in income generation. Activities, which generated cash income more frequent, include non farm income, permanent crops (*i.e.* sale of pineapple, banana, vegetables etc.), seasonal crops, and animal husbandry. Although fish farming had potential to generate income at any time of the year and therefore could fill the gap of income shortage, its income generation surprisingly more-or-less followed a seasonal trend of other income generating activities. The policy maker should give the priority to help the farmers understand about the greater scope of getting frequent income from fish farming by adopting multi stocking and multi harvesting policy. The above information has revealed that the objectives for undertaking fish farming was to a large extent not perceived in comparison with other competing activities. This was attributed to poor production technology, lower adoption rate, harvest strategy and lack of business outlook as hereby explained.

Table 9.3: Average Fish Yield in Respect to Farmers' Objective for Undertaking Fish Farming

	Farmers Objective for Undertaking Fish Farming			
	To Business (12 nos.)	*To Consume (36)*	*To consume and Sale (152)*	*Obtain Status (0)*
Average fish yield	412.05kg/kani	241.58 kg/kani	285.42 kg/kani	–

As depicted in Table 9.3, farmers who realized their objective as fish farming as business point of view intensified the activity and were able to produce on an average 412 kg/kani (1kani=2.5 Acre/0.16ha) of fish that was considered more profitable than other crop production as against the farmers who realized fisheries as only for self consumption as well as selling surplus after consuming in home.

Adoption Index and Adoption Gap of Improved Production Practices of Carp in the Study Area

The extent of adoption scores was measured for nine improved practices *i.e.* pond preparation, liming, manuring the pond, stocking, feeding, water exchange and quality monitoring, use of aerators, health monitoring and management, handling, storage and transportation. The overall adoption index in the study area was 59.74. The production technologies followed by the respondents were described and compared with the recommended technologies. The gap between the existing and recommended technologies was presented in Table 9.4. The Table indicates that non adoption was more in use of aerators (54.25 per cent) to control the oxygen deficiency in the pond. followed by supplementary feeding (51.15 per cent), liming (43.31 per cent), water exchange and quality monitoring (40.37 per cent) and stocking (39.65 per cent). Farmers in general used traditional practices instead of the recommended improved practices may be due to the lack of poor realization of objective, interest and poor financial condition.

Table 9.4: Adoption Gap of the Respondents Regarding Recommended Practices of Improved Production Technology of carp

Sl.No.	Selected Practice	Adoption Gap (per cent)	Rank
1.	Pond preparation	31.41	VIII
2.	Liming	43.31	III
3.	Manuring the pond	26.41	IX
4.	Stocking	39.65	V
5.	Supplementary Feeding	51.15	II
6.	Water exchange and quality monitoring	40.37	IV
7.	Use of aerators	54.25	I
8.	Health monitoring and management	37.28	VII
9.	Handling, storage and transportation	38.25	VI

Preference of Fish Species

In composite fish culture ponds are stocked with compatible indigenous and exotic carps, which have different feeding habits. Therefore, the system gives comparatively a far greater output of fish than those that are stocked with an equal number of either indigenous species or exotic species (Sivasankar *et al.*, 1990). The preferred fish species in the study area has been depicted in Table 9.5. Among the fish species rohu was the most preferred fish species and cultured by all the sampled households. The next preferred species was catla, subsequently then comes mrigal, common carp, silver crap and grass carp respectively. Along with the major carp a few infrequent fish species namely punti, gania, bighead, tilapia, bata, calbasu, pangas, freshwater prawn were found to be cultured by around 22.48 per cent sampled fish farmers.

Table 9.5: Prevalence of Fish Species in the Study Area and their Preference among Farmers

Sl.No.	Fish Species	Per cent Households
1.	Rohu (*Labeo rohita*)	100
2.	Catla (*Catla catla*)	98
3.	Mrigal (*Cirrhinus mrigala*)	82.5
4.	Common Carp (*Cyprinus carpio*)	77
5.	Silver Carp (*Hypophthalmichthys molitrix*)	50
6.	Grass Carp (*Ctenopharyngodon idella*)	20
	Others*	25

*Others: Thai Sarpunti (*Puntius gonionotus*), gania (*Labio genius*), calbasu (*L. calbasu*), pangas (*Pungasius pungasius*), tilapia (*Oreochromosis mssambica*), fresh water prawn etc.

Further study has been done to analyse the criteria which guides the farmers to select the species in their ponds and presented in Table 9.6. The criteria of preference of species shall help us to plan for future research. Yield was found as the most important critera (96 per cent) perceived by the farmers for species preference.

Table 9.6: Species Preference Criteria Perceived by the Farmers

Sl.No.	*Criteria*	*Percentage (per cent)*	*Rank*
1.	Availability of seed	67	VI
2.	Taste	96	I
3.	Low cost of seed	72	V
4.	Yield	85	II
5.	Sale price	82	III
6.	Growth	76	IV
7.	Not damage to pond	54	VII
8.	Resistance to pest and disease	45	VIII

The next important factors are availability of seed (85 per cent), sale price (82 per cent) and growth rate (76 per cent), low cost of seed (72 per cent), taste (67 per cent) etc. among others influence choice of species stocked. Farmers sometimes not preferred common carp to stocked in the pond as the fish sometimes damage the pond. Fish disease was not frequent in the study area therefore, the criteria *i.e.*, resistant to pest and disease was found least preferred criteria for species selection in the study area.

Fish species Mix and Stocking Rate Practiced by the Farmers

The species mix and the stocking rate are two important side of scientific fish culture. Selection of species ratio is generally depends on seed availability, market demand, nutrient status of pond, and management techniques. Average stocking rate and the ratio of different fish species stocked together in the pond along with percentage of sampled households following it has been depicted in Table 9.7. It can be described from the Table that the ideal species mix for composite fish culture *i.e.*, Rohu+Catla+Mrigal+Silver Carp+ Common Carp+ Grass Carp were practiced by only 16 per cent of the sampled households in the average ratio of 24.12: 19.23: 15.45: 16.8: 15.4: 9 with average stocking density of 5107 fingerlings per acre. The species mixture such as Rui Catla and Mrigal were stocked together. Silver carp was stocked later. This could be the reason for avoiding the completion of food between Rui and Silver Carp. No farmers examine the plankton growth by scientific manner as well as toxicity of pond water prior to stocking of the fingerlings. However, sometime by merging the hand in pond water they could identify the natural food availability. A few of farmers followed temperature acclimatization practices of seed before stocking. Some farmers did not include common carp in the fish species mixture to avoid damage of the pond. The adoption of grass carp is lower as comparison to other fish species mainly because the lack of preference for the growers due to unavailability of fingerlings at the major season of stocking. The underlying reason might be due to the fact that breeding season of Grass carp starts late of monsoon in the state and mortality

Table 9.7: Percentage of Fishing Households, Ratio of Fish Species Stocked under different Species Mix and Stocking Rate

Species Mix	*Per cent Households*	*Rohu*	*Catla*	*Mrigal*	*Common Carp*	*Silver Carp*	*Grass Carp*	*Others*	*Stocking Rate (Fingerlings/acre)*
1.	19	32.25	38.26	16.25	13.24	–	–	–	4865
2.	14	34.38	37.8	27.82	–	–	–	–	5060
3.	12	35.25	41.25	–	23.5	–	–	–	3864
4.	20	26.72	23.45	14.28	18.45	17.1	–	–	4770
5.	16	24.12	19.23	15.45	16.8	15.4	9	–	5107
6.	2	34.75	13.25	12.44	16.75	11.98	3.43	7.4	5218
7.	8	24.35	18.6	15.65	22.4	8.5	–	9.5	3287
8.	2.5	62.5	37.5	–	–	–	–	–	2100
9.	2	34.65	–	–	25.24	30.11	–	10	4406
10.	3.5	31.22	21	17.44	25.64	–	-	4.7	4040
11.	2	31.2	25.35	–	–	14.5	18.5	10.45	4860

Others: Sar puti (*Puntius sarana*), Thai Sarpunti (*Puntius gonionotus*), Gania (*Labio genius*), Calbasu (*L. calbasu*), Pangas (*Pungasius pungasius*), Tilapia (*Oreochromosis mssambica*), Fresh water prawn (*Macrobrachium rosenbergii*) etc.

rate between spawn to fingerlings is very high. However, the growth rate of Grass carp in the state was reportedly very high and it can be grown with lower cost feed such as grasses, duckweed, banana leave, carrot leave, azolla, hydrilla, wolfia etc. The consumers demand for grass carp is also increasing day by day. Actually huge demand for live fish has been experienced in Tripura and the preference of species is not the limiting factor at the present situation. Therefore, the productions of grass carp need to be promoted in the state of Tripura by overcoming the barriers in its production.

Realizing Factors Influencing Adoption Process of Fish Farming Technology

Further, the path analysis has been carried out to decompose the total effect of r value into direct, indirect and residual effect of the exogenous variables on the predicted variable *i.e.* extent of adoption of improved production practices of carp culture and depicted in Table 7.9 by incorporating only the significantly correlated factors in analysis. The table revealed that Information management behaviour, training programme attained, potentiality to realize opportunity, innovativeness, knowledge of scientific fish farming occupied the first five ranks in that order. It was further observed that the largest substantial positive indirect effect on adoption of fish farmers was exerted by innovativeness of the fish farmer through their information management behaviour. It was observed that higher information management behaviour combined with innovativeness can lead to higher adoption. It means that when the knowledge of respondents is more through information management, participation in training programme, the adoption of improved fish production practices is also increased. The findings is in the line with Haque(1979) and Goswami (2010). The next important variable is potentiality to realize opportunity. The ability to foresee one's future level of occupational growth leads to successful entrepreneurs and this helps to adoption. Achievement motivation of the respondents motivated their tendency to expand his business and that ultimately change the attitude to adopt different improved practices of fish production. Further, path analysis revealed that training programme attained through Information management behaviour was giving direct effect on the adoption level of farmers. Therefore, extension programmes should concentrate on generating reasoning ability to increase knowledge, improve skills and inculcate required attitude among the fish farmers through training to improve the knowledge and innovativeness. Effort at involving farmers in the mainstream of development must include their training as entrepreneurs. Such programmes, apart from concentrating on entrepreneurial awareness, training and credit reach for farmers must also concentrate upon bringing the new technology within the reach of farmers and acquaint them with the new avenues of entrepreneurship. Through illustrations of success stories the innovativeness of the farmers can be increased. The residual effect being 0.3786, it was to conclude that 37.86 per cent of the variables had been left unexplained. This should further suggest that inclusion of more relevant and contextual variables could have more variations than what have been in present instance.

Table 9.9: Path Coefficients Showing Direct, Indirect and Substantial Indirect Effects of Independent Variables on Adoption Level of Fish Farmers on Improved Practices of Carp Production

Sl.No.	*Variables*	*Correlation Coefficient*	*Direct Effects*	*Rank*	*Total Indirect Effect*	*Rank*	*Variables through which Substantial Indirect Effects are Channeled Through*		
							I	*II*	*III*
X1	Age	–0.264*	–0.094	10	0.472	2	0.1372 X20	0.0504 X17	0.0434 X18
X8	Training programme attained	0.364**	0.298	2	0.216	6	0.1633 X20	0.0681 X11	0.0510 X17
X10	Distance from market	–0.174*	–0.573	9	0.379	4	0.1390 X12	0.0592 X17	0.0540 X18
X11	Average yield	0.173*	0.098	8	0.461	3	0.0601 X20	0.1390 X18	0.0529 X16
X12	Pond size	0.386**	–0.048	11	0.495	1	0.1479 X14	0.0534 X20	0.0494 X16
X13	Potentiality to realize opportunity	0.193**	0.256	3	0.025	11	0.2574 X20	0.1194 X18	0.067X12
X14	Share of fisheries to total income	0.450**	0.107	7	0.320	5	0.1037 X20	0.1237 X18	0.0516X8
X16	Knowledge of scientific fish farming	0.391**	0.167	5	0.166	8	0.2536 X20	0.1294 X8	0.0486 X13
X17	Risk taking behaviour	0.235**	0.130	6	0.209	7	0.1015 X18	0.0701 X17	0.0529 X8
X18	Innovativeness	0.245**	0.241	4	0.098	10	0.2307 X20	0.1268 X8	0.0482 X18
X20	Information management behaviour	0.537**	0.362	1	0.129	9	0.1676 X20	0.1512 X18	0.0587X8

Residual effect: 0.3786.

Conclusion and Recommendation

The objective of this study was to make a thorough investigation on how realizing farmer's objective has influence on the adoption process of fish farming technology in the study area. The results have shown that most farmers adopted fish farming to obtain fish for home consumption as well as for sale. The result further showed that few farmers realized their objectives for adopting fish farming. Consequently, most adopters operated fish farming at low level. Farmer's objectives were not realized due to poor production technology, harvest strategy and nature of the product (*i.e.* farmed fish). Moreover, farmers who realized their objective intensified the activity and were able to produce 412 kg/kani (1kani=2.5 Acre/0.16ha) production that was considered more profitable than other crop production.

The objective of this study was to identify the determinants for adoption of a recommended package of fish farming technology. It is expected that identifying these factors will enable planners in research and extension to advice farmers on aspects, which are compatible with their environment. This will in turn make farmers adopt the recommended package of a fish farming technology. The important conclusion to be drawn from this study is that farmers' access to information about a technology, economic motivation and resources endowment are important in adoption of a recommended package of fish farming technology. The results show that farmers allocate resources to those activities, which meet household needs, more than others. Results reveal that unless the information is communicated clearly which is significant economically, socially to the needs, interests and values of the farmers they cannot utilize the information. The technologies are to be demonstrated in many places of villages with special emphasis on skill components. Farmers should be taken round the nearby progressive farmers' fields and research stations, so that they can accept and utilize the generated technologies.

Realizing farmers' objective has influence on adoption of the technology. This suggests that farmer's objective for undertaking a technology and their priority should be known and taken on board before introducing the technology, and efforts should be made to realize those objectives. For this reason preparation and introduction stages of any technology should involve the farmers. For fish farming to realize farmers objectives, the production technology, harvesting strategy, availability of quality inputs and storage technology need to improve. Production technology including ready availability of quality inputs will ensure that more fish is produced and a farmer friendly harvest strategy will ensure that farmed fish is readily available. Accordingly, improved storage facilities will ensure that farmed fish are properly preserved and can be transported to distant market to make more profit.

References

Anonymous. (2002). Perspective plan for attaining aelf aufficiency in fish production in Tripura 2002-2012 A.D., Department of Fisheries, Government of Tripura.

Batz, F.J., Peter, K.J. and Janssen, W. (1999). The influence of technology characteristics on the rate and speed of adoption. Agr. Econ., 21(2): 121-130.

DoF, Governemnt of Tripura. Fisheries Statistics (2012-13), GoT Retrived from http://http://fisheries.tripura.gov.in/statistics.htm retrieved on 18.02.2015.

FAO, (1996). Aquaculture for local community development and utilization of small water bodies in Southern Africa – Joint Evaluation by Recipient Countries, Donor Government and FAO. FAO, Rome, pp: 115.Edwards *et al.* (1997),

Goswami, B., Ziauddin, G, Datta, S.N.(2010) Adoption behaviour of fish farmers in relation to scientific fish culture practices in West Bengal, 2010, Indian Res. J. Ext. Edu. 10(1) pp:24-28.

Hague, M. (1992). The economic and feasibility of aquaculture in Northern Tanzania. B.A. Dissertation, University of Stirling, Scotland, pp: 40.

Saha, B.(2006). A Study on Livelihood Diversification for Sicio-Economic Development of the Farmers in west Bengal. Unpublished Ph.D. Thesis, IARI, New Delhi.

Sivasankar N, Hugar L B and Umesh K B (1990) An analysis of yield gap and its constraints in inland fish culture of Karnataka State. P 222-224, In: Proceedings of the National Symposium on new horizons in freshwater aquaculture (23-25 January, 1991), ICAR, CIFA.

Temu, A.(1999). Empirical evidence of changes in the coffee markets after liberalization: A Case of Northern Tanzania. Ph.D. Thesis, University of Illinois, Urban - Champaign, pp: 210.

Wetengere, K. (2000). Evaluation of fish farming in Kilosa district, Morogoro region, Tanzania. Unpublished Report. Irish Aid, Kilosa, Morogoro, pp: 35.

Wetengere, K.(2008). Economic factors critical to adoption of fish farming technology. A case of selected villages in Morogoro and Dar es SalaamRegions. Ph.D Thesis, Open University of Tanzania,Dar es Salaam, pp: 242.

Wetengere, K., Osewe, K.and Van Herwaarden, H. (1998). Development of semi-intensive fish farming in Morogoro region, Tanzania. ALCOM Working Paper No. 22. FAO, Harare, pp: 54.

Chapter 10

Pilot Testing, Upscaling and Implementation of e-Velanmai Model of Extension in Tamil Nadu

C. Karthikeyan

Agricultural College and Research Institute, Tamil Nadu Agricultural University, Kudumianmalai – 622 104

Introduction

It is a matter of great concern that there exists wider gap between the proportion of the technologies generated by the public agricultural research system and technologies adopted by the farmers in India. Supporting this fact, the all India value of extension effectiveness index was 47 per cent indicating moderately effective extension mechanism in India (Mishra, 1996). One of the vital reasons for such wide gap is due to the lack of an appropriate extension/technology transfer model in Agriculture that could establish better linkage between Scientists and farmers in order to facilitate effective dissemination of location specific appropriate technologies in a timely manner. Hence there is a need to develop suitable approach for technology transfer in agriculture. Considering this research issue, an attempt was done to conceive an Information and Communication Technology (ICT) based extension model called 'e-Velanmai' (means e-agriculture) for dissemination of farm specific agricultural technologies from the agricultural scientists to the needed farmers in the selected command areas of Tamil Nadu state. This model was subjected to pilot testing and validation to standardize the technology transfer process for achieving maximum effectiveness in solving the farm problems leading to enhanced returns to farmers from agriculture. The pilot experiment was done during July 2007 to March 2011 in three sub basins *viz.*, Palar, Varaghanadhi and Aliyar of Tamil Nadu with the support of the World bank aided TN-IAMWARM project of the Government of Tamil Nadu. Based on the successful results obtained in the performance of the model, World Bank has supported for upscaling the e-Velanmai model of extension in 19 irrigation project command areas of the state during 2011-12 and in 26 sub basins during 2012-13. 'e-Velanmai'project was evaluated periodically and recommended by World Bank experts for implementation as a technology transfer model during 2014-15 in 100 blocks of Tamil Nadu through the extension officials (BTM/ATM) of the Department of Agriculture. This paper presents about the description of the model and its performance in Tamil Nadu.

The specific objectives are

- To upscale the 'e-Velanmai' model of technology transfer to farmers in selected command areas of Tamil Nadu and
- To document the farmer's perception about the ICT based technology transfer model.

Concept of e-Velanmai

It is a combination of personal and ICT based, demand driven, participatory and sustainable extension approach to provide appropriate and timely agro advisory services by scientists to the registered farmers using ICT tools (Internet, Tablet, Mobile Phone etc.) on need and/or regular basis with necessary follow up actions attempted by Field Extension Coordinators (Karthikeyan, 2012).

"e-Velanmai" is described as a combination of personalized and ICT based extension approach due to the presence of human element called Field Coordinator (FC) who handles ICT tools to link farmers and experts for the purpose of technology transfer.

It is also referred as demand driven, participatory extension approach because:

- Farmers paid an annual membership fee (Rs.50-150) based on the farm size owned by them to avail the extension services under 'e-Velanmai'. The nominal fee is collected to ensure the participation of farmers in the system of technology transfer.
- Scientists attended the farmers queries based on their call (demand) or need and hence it is demand driven for technical advice or scientific farming.

It is also believed to be sustainable approach of extension as

- It is envisioned that the farmers could operate and follow the e-Velanmai model of technology transfer by them to access technologies in the long run. The membership fee collected could be utilized to achieve sustainability of the approach. Hence sustainability component is inbuilt in the model.
- Moreover action research on sustainability of the project was attempted and the results indicated that Public-Public model (TNAU and Department

of Agriculture, TN) was sustainable to render cost effective and quality extension service to the farming community.

Components of e-Velanmai

There are six vital components in the technology transfer model (Figure 10.1). They are,

- ☆ *Farmers* who were enrolled as members in the 'e-Velanmai' scheme
- ☆ *Expert team* of Scientists (formed at TNAU)
- ☆ *Field Coordinators* (FC) to develop the capacity of farmers (Extension services)
- ☆ *ICT tools* to link farmers with the experts
- ☆ *Information* about the agricultural problems (data) collected from farmers for advice (Digital images)
- ☆ *Technical messages* (advices) delivered by the experts to solve agricultural problems faced by the farmers and *follow-up* actions on the advices adopted by the farmers.

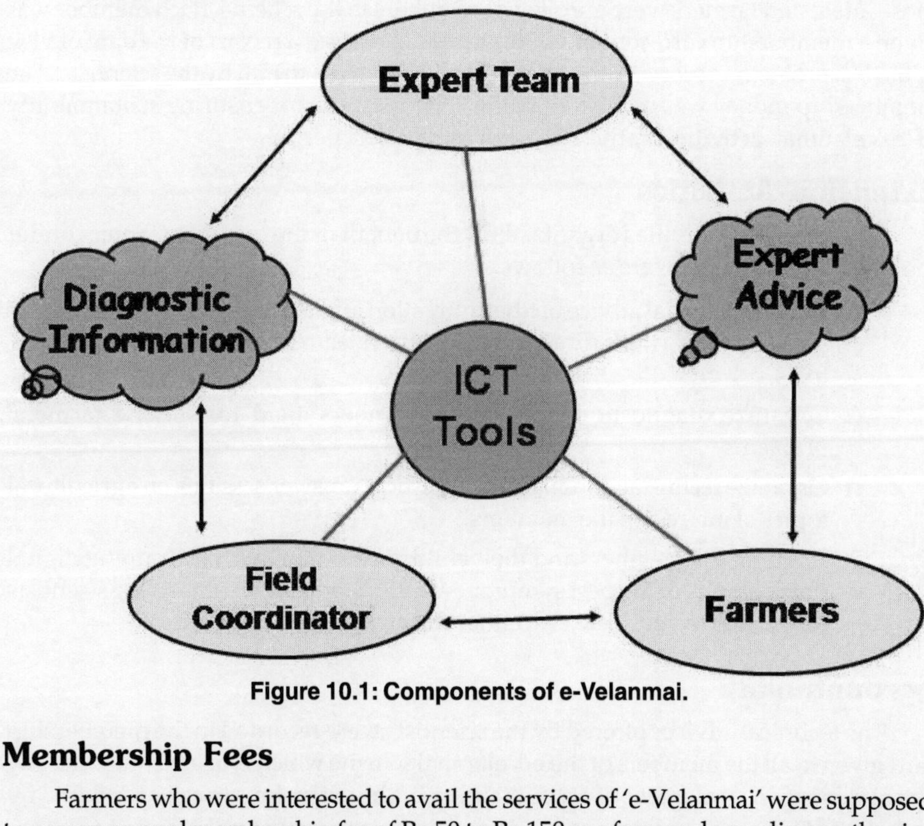

Figure 10.1: Components of e-Velanmai.

Membership Fees

Farmers who were interested to avail the services of 'e-Velanmai' were supposed to pay an annual membership fee of Rs.50 to Rs.150 per farmer depending on the size of their farms (Table 10.1). Fee structure of Rs.50-300 per farmer was decided by the farmers themselves through two brainstorming meetings held at village and WUA level held at Sencherimalai village and Kumarapalayam WUA respectively of Palar sub basin during September 2008. However it was revised as Rs.50-150 during the conduct of the planning workshoponupscalingof e-Velanmai. Fee collection was done mainly to enhance the sustainable participation of farmers in the scientific farming through e-Velanmai model of extension. The fee structure for availing extension services was decided as follows:

Table 10.1: Membership Fee Structure in e-Velanmai

Farm Size (acres)	*Membership Fees Rs./Farmer/Year*
Upto 5	50
5-10	100
>10	150

Those farmers who were interested to avail the ICT based extension advices were considered as members and enrolled in the scheme. Wherever there existed a WUA, its farmers were gathered for an awareness meeting about the e-Velanmai and those interested farmers were enrolled as members in the scheme. Each member was given a membership card containing his farm details and a record of the date of visit of the FC, problems observed and technical advices given to him by the scientists. The membership money was deposited in the WUA account for ensuring sustainability of 'e-Velanmai' activities by the WUA beyond project period.

Extension Activities

The extension activities organized for the benefit of the member farmers under the e-Velanmai scheme were as follows:

- ✰ Timely technical advice on the cultivation and marketing related problems faced by the farmers directly by the expert team from TNAU using the ICT tools.
- ✰ Organize diagnostic field visit to members' field for solving farmers' problems
- ✰ Organize technical seminars and question - answer sessions on agricultural topics of interest to the members.
- ✰ Training the members and their children (Rural Youth) on framing digital photographs of the pest symptoms and uploading the same to the scientists for getting technical advice either through mobile or e-mail.

Accountability

The technical advices offered by the scientists were recorded in the membership card given to all the members of the e-Velanmai scheme which is used to make follow ups and bring in an element of accountability of the extension services rendered by the scientists.

Implementation of the e-Velanmai Scheme (Pilot testing)

Public Model

During July 2007 the scheme was implemented in Palar sub basin by selecting five villages spread over five water users association covering 25 farmers allotting five farmers from each village. One FC was appointed to offer technical advices from the experts to the farmers. The FC visited one village per day to offer extension advice to five farmers thus covering all the 25 farmers in a week. The expert team of Scientists comprising of Entomologist, Soil Scientist, Pathologist, Crop Physiologist, Agronomist and Horticulturist was set up at TNAU, Coimbatore to offer technical advice considering the damaged specimen photographs captured using digital camera and sent to them through www.fileflyer.com website using computer and internet facilities. The advises were given by the Scientists to the FC using mobile phone which were transferred to the farmers in written form by the FC on regular basis without involving any cost to the farmer. Both soil and water sampling were done for all the 25 farms to analyze the soil and water status and the recommendations for the problems were offered considering both the digital images and these sampling results.

Similar model of e-Velanmai was adopted at Aliyar sub basin of Coimbatore district from 1st May 2008 covering 25 farms from five villages drawn from five WUA. The same model of e-Velanmai was replicated at Varaghanadhi sub basin in Villupuram district of Tamil Nadu during 1st July 2008. These two districts were distinctly different in terms of crops, water resource and agroclimatic conditions. However this model was discontinued due to poor response of the farmers to avail the ICT based extension advice offered on cost free basis in the study locale. The experiment gave a lesson that even the quality and timely extension advice when offered to the clients at free of cost on supply driven mode will not motivate the farmers to participate in the extension process (Karthikeyan, 2011b).

For the purpose of ensuring the participation of farmers to benefit from the ICT based extension process, it was thought to alter the implementation strategy from public model to a paid model of e-Velanmai. The paid model was conceived and introduced first at Kumarapalayam WUA of the Palar sub basin from 18th October 2008.

Paid Model of e-Velanmai

In order to enhance the participation of farmers in scientific farming and to have sustainability of the scheme even after the project period, paid model of e-Velanmai was conceived and the same was first introduced at Palar sub basin, followed by Chendur WUA of the Varaghanadhi sub basin from 12th January 2009 and at Mannur WUA of Aliyar sub basin from 1st February 2009.

Under this model, the interested farmers of the selected WUA from the respective sub basins were enrolled as members of the e-Velanmai scheme by collecting the membership fee based on the farm size owned by them. A membership card was given to the farmer and a record of the agricultural problems and advices given by the scientists were noted in the card. These members availed the benefits of e-Velanmai extension activities as mentioned under the public model.

In each of the member's family, at least one person was thoroughly trained in handling the ICT tools for the purpose of framing the crop status images and to access the advices from experts. It was envisioned that the trained individual is expected to capture and send the digital images of the pest damaged symptoms and receive the technical advice from the experts of TNAU. The paid model of e-Velanmai has been upscaled in the following 19 subbasins(Phase I and II) during 2011-2012 (Table 10.2) and 26 sub basins (Phase I to IV) during 2012-2013 (Table 10.3). In each sub basin, one scientist took the charge of implementing the scheme. One FC was

Table 10.2: Upscaling of e-Velanmai (2011-2012)

Sub Basins	
I Phase	*II Phase*
Varaghanadhi	Ponnaiyar
Upper Vellar	Agniyar
South Vellar	Ambuliyar
Pambar	Swethanadhi
Kottakaraiyar	Nichabanadhi
Manimuthar	Kalingalar
Arjunanadhi	Therkar
Aliyar	Upper Gundar
Palar	Koundanyanadhi
	Poiney

Table 10.3: Upscaling of e-Velanmai (2012-2013)

Sub Basins	
I – PHASE	Nallavur
Varaghanadhi	Ongur
Upper Vellar	Thurinjalar
South Vellar	Gadilam
Kottakaraiyar	Gridhamal
Manimuthar	Uthirakosamangaiyar
PAP - Palar	Deviyar
II – PHASE	Hanumanadhi
Ponnaiyar	Kambainallur
Agniyar	**IV – PHASE**
Swethanadhi	Cooum
Koundinyanadhi	Adayar
III – PHASE	CheyyarKillyar
Araniyar	Paralaiaru
Kosasthalaiyar	Amaravathi

appointed to link the farmers and scientists for the purpose of technology transfer. An expert team comprising of crop protection and crop management was identified in either a KVK/Research station located in each of the sub basins to offer scientific advices to farmers.

Operational Schedule for Technology Transfer

The operational procedure in the technology transfer process (Figure 10.2) was initiated by any farmer who is in need of technical advice in the sub basin. The farmer calls the FC over his mobile and the FC goes to the farmer's field and examines his crop status. In case the farmer is in need of decision based advices such as suitable variety, marketing decisions, spacing etc., then the advices were given on the spot by the FC by referring to the TNAU agri portal. In the case of a problem based query raised by the farmer say for example the management of an insect pest, disease, nutritional deficiency, weeds etc., the advices will be given to the farmer based on a digital photo document captured by the FC and sent to the experts using internet. The images were coded using the farmer's name and identity number. The experts diagnosed the problem photos and offered solutions to the respective cases. These recommendations were downloaded and recorded in the membership card given to the farmers respectively.

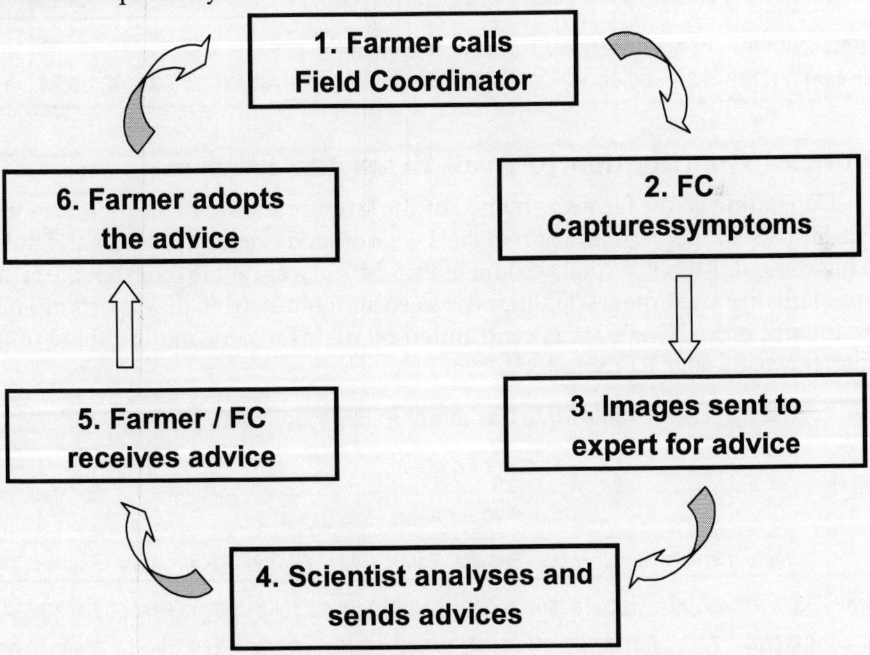

Figure 10.2: Stages in Technology Transfer Process in e-Velanmai (2007-13).

Results and Discussion

The results on the experimentation (Pilot test +upscaling) of the scheme and the lessons learnt are presented as follows:

Membership Status in e-Velanmai

Farmers were enrolled in the scheme as members by collecting a membership fee of Rs.50 to 150 per farmer based on their farm size. About 10507 farmers had joined the scheme (Table 10.4) by paying the prescribed annual membership fees. The gender wise distributions among the members revealed that 89.76 per cent were male and the rest were female. The farm size wise distribution of the members indicated that a majority of small and marginal farmers (80.60 per cent) who operated < 5 ac land holding had participated largely as compared to medium and big farmers. The fact that more than three-fourth proportions of farmers in the state were small farmers could be attributed for the result obtained. The results suggest that farmers irrespective of their land holding had expressed their willingness to pay behavior towards availing quality and timely extension advices (using ICT tools) to adopt scientific practices in their farms.

Table 10.4: Membership Enrollment in e-Velanmai

26 Sub Basin	*No. of Farmers Enrolled*									*Total No. of Members Enrolled*		
	<5 ac			*5-10 ac*			*>10 ac*					
	Male	*Female*	*Total*	*Male*	*Female*	*Total*	*Male*	*Female*	*Total*	*Male*	*Female*	*Total*
Total	7490	979	8469	1814	93	1907	126	4	130	9431	1076	10507
Per cent	71.29	9.32	80.60	17.26	0.89	18.15	1.20	0.04	1.24	89.76	10.24	100

Farmers' Contribution to Avail Extension Services

Depending on the farm size owned by the farmers, they paid the membership fee of Rs.50 to Rs.150 per farmer. The members were offered extension advices and training on handling ICT tools. A total amount of Rs.6,54,700 was mobilized from the farmers to offer quality and timely scientific advices in agriculture from the experts of TNAU. An amount of Rs.5,94,050 was contributed by male farmers and the share of farm women was Rs.60,650 (Table 10.5).

Table 10.5: Membership Fees Collected

Sub Basin	*No. of Farmers Enrolled*									*Total No. of Members Enrolled*		
	<5 ac			*5-10 ac*			*>10 ac*					
	Male	*Female*	*Total*	*Male*	*Female*	*Total*	*Male*	*Female*	*Total*	*Male*	*Female*	*Total*
Total	374900	49300	424200	189500	10300	199800	29650	1050	30700	594050	60650	654700
Per cent	57.26	7.53	64.79	28.94	1.57	30.52	4.53	0.16	4.69	90.74	9.26	100

This amount was deposited in rural banks by creating a separate account in the name of the social organization/WUA maintained as joint account to be operated by the leader of the social organization and the scientist-in-charge of the sub basin. The money is proposed to be used to avail extension advices/agril. information from the experts/media sustainably after the closure of the scheme in the sub basin.

Awareness Meetings and Seminars

Trainings include the awareness meetings and technical seminars organized for the benefit of the farming community at village/WUA level in each of the selectedsub basins. The farmers were also trained to handle the ICT tools such as framing the crop status images, uploading the images to the scientist and accessing the scientific advices using computer and internet to solve farming problems. The Table 10.6 results reveal that 9789 trainings were organized in all the selectedsub basins covering all the members of e-Velanmai. These trainings were benefitted by 13919 farmers.

Table 10.6: Total Number of Trainings Conducted

Number of WUA/ Social Groups	*Total Number of Trainings/ Meetings*	*No. Trained*
430	9789	13919

Capacity Building on Handling ICT Tools

The farmers were trained to handle digital camera to capture crop status images that can express the pest attacked symptoms clearly. Trainings were also organised to the farmers for the purpose of accessing scientific advices through internet. All the 10507 members of e-Velanmai were trained in handling either or both the ICT tools. Among this trained category of farmers, a majority of them were trained in handling a camera (74.82 per cent) and about 25.18 per cent were trained in internet media. The category of trained farmers revealed that more than half of the trained group were farmers' wards (80.24 per cent) followed by youngfarmers (19.76 per cent) Experiences reveal that the children or the youth belonging to the farmers' family had expressed relatively higher level of interest and favourable attitude to learn how to handle ICT tools as compared to the farmers/farm women. It was intended to train at least one member in each of the farmer's family so that they can help their family members in accessing the advices from the experts in future.

Agricultural Extension Advisory Services

Experiences on answering the queries raised by the farmers revealed that there existed two types namely (i) Problem based queries and (ii) Decision based queries. Problem based queries were answered with the help of the crop status images (Eg. Pest damaged symptoms) whereas the decision based queries (Eg. Marketing decisions) were answered on the spot using the TNAU agriportal website and other technical data base in agriculture. These queries were also analysed according to the subject matter it represented. The results reveal that about 20,259 queries were tackled by offering suitable technical advices to the farmers. Among these queries, a majority come under the plant protection subject (54.22 per cent) followed by crop management (34.36 per cent) and other subjects (11.42 per cent) such as Marketing, Post harvest technologies, Agricultural Engineering etc. It is a well known fact that the presence of an insect pest or disease or lack ofmicro and macro essential nutrients in the crop will express damage symptoms and hence these problems were visible and makes the farmer to get concerned about the problem. This might have influenced the farmers to raise more problem based queries (73.97 per cent) focussed on crop protection aspects followed by management decision based queries (26.03 per cent). *It is proved that an agricultural expert can offer effective technical advices to farmers based on the digital images of the crop status, even without visiting the farmers field in person* (Table 10.7).

Table 10.7: Technologies Transferred to Farmers

Year	Subject-wise No. of Advices given								Total No. of Advices given		Grand Total
	Pathology	Entomology	Physiology	Soil Science	Agronomy		Marketing	Others			
	PB	PB	PB	DB	PB	DB	DB	DB	PB	DB	
Advices	5643	5343	1577	496	764	4117	1301	1028	14985	5274	20259
Per cent	27.85	26.37	7.78	2.49	3.77	20.32	6.42	5.07	73.97	26.03	100

Adoption Status of the Advices Offered to Farmers

The farmers' queries were answered by the experts in the case of problem based queries and the decision based queries were answered on the spot by the FC or after getting the advice from the experts. These advices were adopted at farm level by an overwhelming majority of the farmers. The ratio between the number of farmers queried to adopted was found to be 84.61per cent (Table 10.8). The success rate of the adopted farmers was found to be 91.10per cent and the remaining seems to be landed in failure due to various reasons such as shortage of labourers for adoption of recommended practices, lack of conviction of farmers over the recommended practices and high cost of recommended inputs.

Table 10.8: Adoption Status

26 Sub basins	Number of Farmers Queried	Number of Farmers Adopted	*Success Rate (per cent)
Total	20259	17142	91.10

Rate of adoption 84.61 per cent.

* Success rate due to the adoption of advices among the adopted farmers.

Reasons for Non-Adoption of the Recommended Advices

The problems confronted by the farmers were identified either by the Scientist or the FC and the appropriate advices were recommended to the farmers in time within 3 hours. However some farmers did not adopt the advices in their farm due to various reasons. Hence a survey was collected with 500 randomly selected samples from all the selected sub basins and the results are given in the Table 10.9.

The findings on the reasons for non-adoption of the recommended advices revealed that a majority of the farmers (23.60 per cent) felt that they were not able to procure the recommended agro inputs from the local input shops. Shortage of labour to perform various agricultural operations in the village was also identified as one of the reason for non-adoption (20.60). The other reasons for non-adoption of the advices include, high cost of the recommended inputs (13.60 per cent) such as Methomyl, Nativo etc., which was sold for Rs.1000 to 6000 per kg. Lack of conviction of farmers for want of observable results in the field immediately after adoption of the practices (17 per cent), resistance to change attitude of farmers (7 per cent) for adoption of new technologies.

Table 9. Reasons for non-adoption of the recommended advices

Sl.No.	*Reasons for Non-Adoption of Advices*	*No. of Farmers* (N=500)*	*Percentage*
1.	Non-availability of prescribed pesticides in local area	118	23.60
2.	Shortage of labourers for adoption	103	20.60
3.	Lack of conviction of farmers for want of immediate observation of results	85	17.00
4.	High cost of recommended inputs	68	13.60
5.	Resistance to change attitude of farmers	35	7.00

* Multiple responses recorded.

Reasons for Failure after Adoption of the Advices

The technologies advised to the farmers were not adopted scientifically as per the correct dose and specifications recommended. Hence 9 per cent of the farmers reported failure in their farms (Table 10.10). Agricultural operations have to be done in time to achieve a successful crop. However depending upon the availability of the labour in the village as well as the sprayer availability on hire basis from the owner farmers, the spraying of pesticides was undertaken by the farmers which had not favoured the farmers for a successful crop. Hence this had resulted in failure as reported by 7.6 per cent of the farmers. Some farmers (4.4 per cent) opined that the input shop keepers interfered in altering the suggestions of the experts and attempted to recommend higher dosage of inputs/alternate inputs thus preventing the farmers to adopt as per the advices of the experts which had not yielded good results in the

Table 10.10: Reasons for Failure after Adoption

Sl.No.	*Reasons for Failure After Adoption of Advices*	*No. of Farmers* (N=500)*	*Percentage*
1.	Improper adoption against recommendation	45	9.00
2.	Spraying based on the labour/sprayer availability	38	7.60
3.	Interference of the private dealer in adoption of the recommendation	22	4.40
4.	Adopting the recommendation at inappropriate stage of the crop	18	3.60
5.	Urgency in sowing operation ignoring the seed treatment technology	18	3.60

farm. Few farmers (3.6 per cent) received the recommendation in time but attempted to adopt latter thus missing the appropriate growth stage of the crop leading to failure. Other group of farmers had reported that they had ignored the recommended seed treatment technology in paddy due to sudden onset of monsoon in their village thus landing in the failure of the crop.

Benefits/Advantages Realized by Farmers Due to e-Velanmai Model of Extension

The farmers were enquired about the benefits they realized due to the adoption of technical advices received through the e-Velanmai method of extension. The results are presented in the Table 10.11.

Table 10.11: Benefits/Advantages Realized by Farmers

Sl.No.	Benefits Realized	No. of Farmers* (N=500)	Percentage
1.	Timely advices were received at individual's farm level/door step	455	91
2.	Saved input cost to manage/protect a crop	375	75
3.	Reduced crop loss/obtained higher yield and earned more income	365	73
4.	Precise and appropriate recommendations adopted to tackle problems	310	62
5.	Dependence over input dealers for advices decreased	240	48
6.	Adopted improved/scientific methods to manage the pest and diseases	180	36
7.	Gained awareness/knowledge about the management of various pests/symptoms/and fertilizer dosage	165	33
8.	Use of biofertilizers, biocontrol methods for different crops were learnt through FC	125	25
9.	Awareness/acquired skill to operate ICT tools like camera and internet for further contact with experts	125	25
10.	Functioning of social group through e-velanmai scheme	75	15

* Multiple responses recorded.

It was observed that an overwhelming proportion of farmers felt that they were able to receive the advices timely on the same day when they raised the query to the field coordinator of the project. They were also happy to enjoy the benefit of access to the recommendations either at their door steps/farm level itself. Nearly three-fourth of the beneficiaries opined that they were able to achieve input use efficiency *i.e.,* saved money on investment over inputs use to manage the crop production or protect the crop from various pest and diseases. Almost an equal proportion of farmers reduced their crop loss due to the adoption of the recommended management practices resulting in achievement of higher yield and income. More than half of the beneficiaries reported that they were able to adopt the precise and appropriate recommendations to tackle various problems confronted in their farm.

Prior to the introduction of e-Velanmai project in the village, nearly half of the farmers stated that they depended on the input shop keepers/dealers available in their locale for crop protection advices. This situation had changed for them after the introduction of e-Velanmai as they were able to access all advices related to agriculture at their farm gate itself. About one-third of the farmers reported that they were able to adopt improved management/crop protection practices. In other words, one-third of the farmers stated that they had come to know about various pests, their symptoms and management technologies in agriculture.

About one-fourth of the beneficiaries expressed that they had learnt about the use of bio-fertilizers and bio-control methods in crop protection from the field coordinators as such technologies were recommended to the farmers. Farmers also reported that they had acquired skill to operate the ICT tools such as digital camera to capture crop status images or internet to access the recommendations from the experts of the project. Few farmers had opined that they gained social recognition and support to share agricultural information among a group of farmers due to the functioning of social groups formed under the e-Velanmai project at their village level.

It could be inferred that farmers had endorsed several economic, personal and social benefits/advantages due to their participation in the e-Velanmai project to access and adopt scientific advices to tackle their farm related problems.

Problems Perceived by Farmers to Access/Adopt Advices from e-Velanmai

The farmers were asked about those problems faced/expected in future by them to follow the e-Velanmai mode of accessing the recommendations/adopting the advices from the experts. The responses were gathered and presented in the following Table 10.12.

Table 10.12: Problems Perceived by Farmers to Access/Adopt Advices from e-Velanmai

Sl.No.	*Problems*	*No. of Farmers (N=500)*	*Percentage*
1.	Faced problems in the use of ICT tools for access to advice by themselves	165	33
2.	Non-availability of ICT tools	150	30
3.	Poor internet connectivity in rural areas	110	22
4.	Unexpected rainfall wash away the sprayed pesticides	100	20
5.	Pesticide and fungicide names labeled in English	60	12

Although the farmers in e-Velanmai project received appropriate advices from experts few farmers had expressed some problems as discussed here. About one-third of the farmers lack confidence and faced the problem of handling ICT tools namely digital camera, internet and computer to access the advices from the experts by themselves without the help of the field coordinator. About 30 per cent of the farmers although they had knowledge to operate the ICT tools, they expressed that it was not possible for them to adopt e-Velanmai model to access advices by themselves due to the non availability of ICT tools with them.

Poor internet connectivity existing in the villages was expressed as one of the problems to access advices by about 22 per cent of the farmers who were either confident or supported by their trained family members to access the recommendations from the experts in future.

One-fifth of the beneficiaries reported that even though they were successful in adopting the recommendations given by the experts, they landed up in failure due to the unexpected rainfall that occurred after the spray of the recommended pesticides in the field. About one-tenth of the farmers expressed their inability to understand whether the same pesticides as prescribed by the experts were issued by the input shops. This is due to the illiteracy condition of the farmers which limited them to read the name of the input that is labeled in English version.

Any innovation will certainly have its own problems to perform successfully at field level and e-Velanmai model of technology transfer is not an exception. However the model performed effectively at field level and the problems listed by the farmers are not the lacunae existing in the method of extension. Improvement of economic conditions and change of the mindset or attitude of farmers to learn and use to handle ICT tools could overcome the problems in near future.

Experiments on Sustainability of 'e-Velanmai'

The sustainability of offering the extension advisory services through 'e-Velanmai' approach even after withdrawal of support from TN-IAMWARM project was studied. Three models to sustain e-Velanmai was experimented and the results are presented as follows.

Model – I: Involving Rural Youth as FC

The option of engaging the interested rural youth to assume the role of Field Coordinator for the purpose of technology transfer was field tested at Palar sub basin. Mr.Sakthivel from Senjerimalai village had shown interest to perform the role of field coordinator. He was promised to be paid Rs.10 for every case he dealt with for providing advisory services. He got trained to handle ICT tools for technology transfer purpose. The ICT tools were provided to him from the project. From the researcher's assessment, he was not capable enough to offer appropriate advices to the farmers, as he lacked the technical knowledge which is essential to perform appropriate advisory services to farmers. From his personal view, he felt that on an average maximum of five cases per day were attended by the youth. Hence he demanded an amount of Rs. 300 per day as salary from the project. Four such volunteers from different villages of Palar sub basin were approached and all of them were of the same attitude of earning a fixed and assured income from the project. Considering the attitude of the rural youth, it was not judged as a sustainable option and hence the idea of involving the rural youth did not work out fruitfully to sustain the project.

Model – II: Public – Private Partnership

The Tata Consultancy Services (TCS), Chennai was approached to collaborate with the project and offer the advisory services for the farming community of

Villupuram district and latter to other areas in Tamil Nadu. TCS gave a proposal regarding the modalities and means by which they can venture into the project. As per the proposal, TCS has described that they could collaborate with TNAU under certain conditions.

The subscribing farmer will have to pay annual fees to avail the extension services. In the 1st, 2nd and 3rd and 4th year the fees would be Rs.1000, Rs.1200, Rs.1300 and Rs.1400 respectively. In the first year the fees would be 100 per cent subsidized.

Farmer will have to bear the cost of GPRS charges. In the first year TN-IAMWARM project would bear the GPRS cost. The cost could differ from one telecom operator to other. BSNL provides free data download on certain recharge packages but the quality is poor. The average per farmer per year cost would be around Rs.1200.

The willingness of farmers towards the TCS proposal was enquired with the farmers of Varahanadhi, Thurinjalar, Ongur and Nallavur sub basins. None of the farmers in all the four sub basins were willing to accept the idea as it involved huge cost sharing from the farming community to access ICT enabled extension services. The researchers also felt that the proposal given by TCS is a typical business oriented model rather than service oriented one and hence there is no point in proceeding further with TCS to sustain the project. In case if TCS had considered offering the extension services to farmers under the banner of Corporate Social Responsibility initiatives, then the project would be feasible and sustainable in the long run. Hence the PPP venture between TNAU and TCS was proved to be a futile attempt.

The Indian Tobacco Company Ltd. is already providing ICT enabled agricultural extension services to the farmers in selected areas of Sivaganga district. They have already established ICT infrastructure in selected villages and employed field staff to offer extension advisory services to farmers. They were willing to collaborate with TNAU for obtaining the technical advices from scientists and transfer the same to the farmers. However the scope of operation is limited to the selected areas of the district only. They were not willing to extend their area of operation to cover TN-IAMWARM Sub basins of the district. Hence the project could not be extended sustainably either in sub basins of TN-IAMWARM or other parts of Tamil Nadu.

Model – III: Public-Public Partnership

It was felt that the state department of agriculture which is holding the major responsibility to transfer agricultural technologies to farmers in the state could be an option to sustain the project. Accordingly, the Joint Director of Agriculture, Coimbatore district and the Deputy Director of Agriculture (DPAP), Coimbatore was approached to instruct the Block Technology Manager (BTM) working under ATMA of one block coming under Palar sub basin namely the Sultanpet block to enact the role of Field Coordinator (FC) and offer ICT enabled extension advices to the farmers. The officials accepted and gave oral instructions to the BTM for offering the extension advices to farmers using the already available data base regarding the crop status and recommendations offered to the Palar sub basin farmers. An awareness meeting to the farmers about the involvement of the BTM as FC was held on 26th July 2011 at Senjerimalai village. The meeting was conducted by the top Officials of the Agriculture department and TNAU. It was attended by the members and office bearers of four

WUA. Farmers accepted the idea to obtain the e-Velanmai mode of technology transfer through BTM instead of FC engaged in the project. Farmers were willing to pay and access the ICT enabled extension services. All the BTMs of Coimbatore district were trained on 21st June 2012 at TNAU, Coimbatore to implement e-Velanmai.

It was found that the BTM of Sultanpet block is capable to transfer appropriate technologies to farmers using the existing data base on crop status images and advices developed under the project (Karthikeyan, 2012). It was also understood that BTM/SMS could be involved effectively to offer ICT enabled agricultural extension services to farmers and the project could be operated sustainably in Tamil Nadu. However it is felt that suitable orders from the Government of Tamil Nadu is needed to fix the role of providing ICT enabled services by the BTM/SMS of ATMA as one of the extension responsibility in addition to other existing responsibilities being attended by the BTM/SMS.

Considering the socio-economic scenario of farmers in Tamil Nadu, *it is stated that Public – Public partnership could be a viable and feasible option to sustain e-Velanmai project in Tamil Nadu,* for which suitable orders are also essential from the Government of Tamil Nadu.

Lessons Learnt

The lessons learnt from the action research project is listed as follows:

1. Appropriate farm specific and timely agricultural extension advices can be offered by the agricultural experts to solve the farmer's problems based on the crop status images sent to by field extension staff or farmers from the field using ICT tools.
2. The limitations of physical distance that exist between the expert and farmer besides inadequate availability of transportation facilities for performing extension advisory services could be overcome by e-Velanmai method of technology transfer.
3. Farmers will come forward to pay a nominal fee based on their farm size for accessing quality and timely extension advices from agricultural experts.
4. Most of the farmers were willing to pay for accessing agricultural advices through ICT mode like e-Velanmai. Hence, farmers were ready to pay for accessing quality advices which are rendered timely by the experts.
5. Farmers across gender and farm size had shown interest and participated in the ICT enabled extension process for adoption of scientific farming.
6. The upscaling of e-Velanmai project is feasible, workable and replicable in the irrigation project command areas of Tamil Nadu.
7. It is possible to train at least one member of the farmers' family to handle ICT tools to access scientific recommendations from the experts.
8. It is possible to sustain the ICT enabled personalized extension approach by involving the BTM/ATM of ATMA as FC for technology transfer to farmers.

Suggestions

1. e-Velanmai combines both personal approach and ICT approach. So it was made possible to offer location specific services to the farmers. Considering the existing extension manpower of the state department of agriculture and the ICT usage level of farmers, a combination of both personal and ICT enabled extension approach is suggested.
2. In order to sustain the project activities in the long run, Public-Private-Partnership arrangements with the IT based companies, NGOs and Agro input manufacturing industries who are interested to extend ICT enabled farming advisory services as a part under Corporate Social Responsibility initiatives need to be explored. Alternatively, the Department of Agriculture of the Government of Tamil Nadu can adopt the e-Velanmai model as one of the effective means to provide technical advises to farmers and enhance farm productivity at state level.
3. It is suggested to train all the Assistant Agricultural Officers, Block Technology Manager (BTM) and Assistant Technology Managers (ATM) of ATMA scheme working in the Department of the Agriculture and allied departments so that they can assume the role of ICT based technology transfer service as one of their responsibility.

Adoption and Implementation of e-Velanmai by the State Department of Agriculture (2015-16)

The World Bank experts took the evaluation of e-Velanmai project periodically and recommended the extension approach for adoption by the State Department of Agriculture in 100 Blocks of 13 districts in Tamil Nadu state. Suitable Government orders (Vide G.O.4 (D). No. 11 Dt.15.05.2014 Agriculture (WD2) Department of the Agril. Production Commissioner and Secretary). An amount of Rs.50 Lakhs was sanctioned to execute the project jointly by the Tamil Nadu Agricultural University and State Department of Agriculture during 2015-16.

e-Velanmai Extension Approach Implemented during 2015-16

An android app. for performing e-Velanmaiadvisory services was created under the project and integrated with the Farm Crop Management System (FCMS) software of the Department of Agriculture. This e-Velanmai-FCMS app can be loaded in mobile/ tablets with android platform. The extension officers, specifically the Block Technology Managers (BTM) and Assistant Technology Managers (ATM) working under ATMA project were engaged to perform the extension advisory services to the farmers in the 100 Blocks. The extension officials were provided with tablet loaded with e-Velanmai-FCMS android app. About 200 Assistant Directors of Agriculture and Agricultural Officers working in the 100 blocks were trained on using the app to execute technology transfer services to the farming community during June – July of 2015. These master trainers will in turn train the BTM/ATMs on e-Velanmai method of extension who are working under their control in the 100 blocks of Tamil Nadu.

Stages in Technology Transfer Using e-Velanmai Android App

The BTM/ATM of each block will visit the farmer's field in their block and capture the problematic crop status images along with the farmer's mobile number. These images will be sent to the Experts at Level I (Agricultural Officer of the Block) by opening the log in page of the e-Velanmai android app. using the pass word and user name assigned for a particular block. The experts at Level I will open the advisory page of the e-Velanmai-FCMS app. in the tablet to diagnose the symptoms of the pest and send the SMS - advices in Tamil version to the farmer whose crop is affected by the pest. In case, if he is unable to identify the pest/give a recommendation he is at freedom to export the crop status image to the expert at Level II (TNAU Scientist) seeking an appropriate advice for the farmer. The team of experts at Level II would open the e-Velanmai-FCMS app. to offer recommendation to the farmer for tackling the pest in his crop. The advices will be prepared in Tamil version and the same will be sent to the farmer as well as the extension officer who has sent the image to the scientist. The SMS services offered at free of cost by the Kisan SMS portal of the GOI is being accessed for sending the advices in Tamil version to the farmers through this project. Later during subsequent visits the BTM/ATM would visit the farmer's field to check the adoption status of the farmers.

Application of the Extension Model to other States in India

The e-Velanmai approach of extension has been successfully pilot tested, upscaled and further implemented for adoption by the extension officials of the state Department of Agriculture in Tamil Nadu. This extension approach is found to be reliable and valid one to disseminate cost-effective and timely advisory services to the needed farmers over mobile phones in the form of SMS advices and followed by a technical person (BTM/ATM) to clarify and guide the adoption process. This does not involve huge investment and technical manpower to execute at field level. The basic infrastructural requirements for implementing this project are mobile phone, tablet with internet to be owned by the experts. Hence, this extension model can be applied for adoption/implementation in all the states of India as well as all developing countries of our world.

References

Abu, G.A., T. E. Taangahar and I. D. Ekpebu. 2011. Proximate determinants of farmers WTP (willingness to pay) for soil management information service in Benue State, Nigeria. African Journal of Agricultural Research, Vol. 6(17): pp 4057-4064.

Berlo, D. 1960. The process of communication: An Introduction to theory and practice, New York: Holt, Rinehort and winston.

Karthikeyan, C. 2011a. "e-Velanmai: An Information and Communication Technology (ICT) enabled agro-technology transfer model". Paper presented in the International Conference on 'Innovative approaches for agricultural knowledge management - global extension experiences' held during 9-12, November 2011 at NASC Complex, New Delhi.

Karthikeyan, C. 2011b. "e-Velanmai: Pilot testing and upscaling of ICT enabled agricultural extension model". Lead paper published in the Compendium of the National Seminar on "Innovations in farming systems research and extension for inclusive development", (Edited by Sudeepkumar,N.K., P.Thilakar, N.V.Rajkumar, T.Senthilkumar, P.Thirunavukarasu) held during 24-25, November 2011 at Tamil Nadu Veterinary and Animal Sciences University, Chennai.

Karthikeyan, C. 2012. "e-Velanmai" – An ICT Enabled Agricultural Extension Model. International J. of Extension Education, Vol.8: pp 24-30

Karthikeyan, C., N. Ajjan, S.Chellamuthu, and B.J.Pandian, 2012. "e-Velanmai: Action research on ICT enabled technology transfer model in Agriculture" Paper published in the Proceedings of the International Symposium on "100 years of Rice Science and looking beyond" under the theme 'Rice science for economic prosperity and nutritional security' Vol. I Crop Improvement, Rice Mechanisation and Rice Policy' (Edited by SreeRangasamy, S.R., K.Thiyagarajan, S.Robin, R.Rabindran, S.Suresh, S.Manonmani, S.Rajeswari, P.Jeyaprakash, V.Ravichandran, S.Radhamani and R.Pushpam) held during 9-12, January 2012 at Tamil Nadu Agricultural University, Coimbatore.

Karthikeyan, C., "e-Velanmai" An ICT enabled Agricultural Extension model for scientific sugarcane cultivation under Public-Private-Partnership venture. Paper published in the Proceedings of the International Symposium on New paradigms in sugarcane research held during 15-18, October 2012, Organized by Society for Sugarcane Research and Development, Coimbatore and Sugarcane Breeding Institute, Coimbatore.

Mishra, D. C., 1996. "Agricultural extension effectiveness in India. A cross sectional analysis", Indian Journal of Extension Education, Vol. 32, No.1-4.

Chapter 11

Institutional Innovations for Marketing of Farm Produce: KVK Experiences

H.K. De and A.K. Singh

ICAR-Zonal Project Directorate, Zone II,
Kolkata – 700 097, West Bengal

State agricultural Universities and research institutes under Indian Council of Agricultural Research (ICAR) are organizing 'Innovators meet' with the objective of documenting farmers' innovations. Several such innovations have been documented in the field of agriculture and allied sectors. These innovations provide opportunities to the researchers to work upon and improvise. Similarly, in the field of marketing of farm produces there are several such novel approaches initiated by various stakeholders. Krishi Vigyan Kendras, working in the district level are the frontline extension system providing need based technical support to farming community. Few KVKs have also initiated new approaches of bringing buyer and seller together. Some such cases holds immense potential for replication *viz.*, sales counters at KVK Ranchi; processed food products being sold to CRPF (Hazaribag); Lac marketing in Palamu; Mushroom in mid day meal scheme (Nalanda); Satellite centre of KVK (West Midnapore) etc. The catalytic role played by KVKs in marketing farm produces is worth documenting. The out-of-box solutions developed at the grass root level needs upscaling by forging forward linkages. A few selected institutional innovations in market linkage facilitated by Krishi Vigyan Kendras are discussed in the article.

Introduction

The terms "creativity" and "innovation" are often used interchangeably.

However they both have distinct connotations. Creativity is about coming up with ideas while innovation is about "bringing ideas to life." While individuals may display creativity, innovation occurs in the organizational context only, by bringing creative ideas to life. Innovation is linked to performance and growth through improvements in efficiency, productivity and quality. Innovation by businesses is achieved in many ways; through formal research and development for "breakthrough innovations" as well as through less formal on-the-job modifications of practice-such as through exchange, personal experience and by many other routes.(http://www.bia.ca/what-is-innovation.htm)

The term innovation refers to the understanding and use of a new idea, practice or method, which replaces something that an individual or organization has been using so far. Innovation can be conceptualized as a "product" or "end result" usually measured as adoption rates by some authors such as Rogers (1962, 1995), while other authors conceptualize innovation as a process; for example, the process of generating new knowledge and applying it in a productive way (Hall *et al.*, 2003, 2004).

New approaches to innovation have their roots in structural adjustment programmes implemented during 1991 (Oscar *et al.*, 2013). This resulted in shrinking government spending on agricultural research and extension services. A number of new stakeholders (NGOs, private sector, farmer organizations, local governments, etc.) started to contribute and, in some cases, take the lead in promoting agricultural innovations during the 1980s and 1990s (Bebbington *et al.*, 1993; Umali and Schwartz, 1994; Ameur, 1994; Ortiz, 2006). The appearance of new stakeholders involved in agricultural innovation, and the realization of their importance among donors and scholars, highlighted the need to improve linkage mechanisms for technological innovation, for example, through participatory research (Chambers *et al.*, 1989), or for improving the share of information and knowledge (Kaimowitz *et al.*, 1990).

In several cases innovations started to occur as a result of a combination of the comparative advantages of public and private stakeholders, among which research organizations were just one among several players. As the changes occurred, scholars started to propose new theories aiming at explaining how multiple stakeholders interact and innovate to solve common problems. The agricultural innovation system approach (Hall *et al.*, 2003, 2004; Hall, 2009) was initially used in the private sector in an attempt to explain and promote innovation.

ICAR (Indian Council of Agricultural Research) Institutions and SAUs are organizing Farm Innovators' meet. Selected innovators are invited by the Institutions to share their way of farming to scientists and other stakeholders. These meets act as an excellent platform for the stakeholders who get first-hand information about grassroots level innovations besides providing scientists direct feedback on the performance of the technologies. ICAR's publication "Farm Innovators 2010' compiled 139 innovations from across the country ranging from crop, water management, livestock and fisheries etc. the documents are like goldmines for scientists. They get to know personal experiences of farmers, fine-tuning of recommended practices done at the farm level by farmers, and other modifications that have become successful. It also opens up opportunity for researchers to improvise farm innovations.

Table 11.1: Selected Institutional Innovations in Marketing

Sl.No.	Name of KVK	Crop/ Enterprises	Linkage Developed	Remarks
1.	Divyayan KVK, Ranchi	Honey and other products	Sale counter at KVK and Railway station	KVK Ranchi has a well established sale counter for sale of training products, farmers products etc. which is functioning for more than 30 years. Major products, prepared by trainees, sold from the sale counter are honey, vermicompost, eggs, birds, flowers and fruit seedling and sapling, seasonal vegetables, tomato sauce, nutritive laddu, mushroom, spawn, fish, cow milk etc. KVK collects products from the trained farmers, testing is done in specialized laboratory and after testing good quality products are procured from the farmers and is sold from its sales counter. KVK also runs a counter at Ranchi rail station. During 2013-14 KVK Ranchi sold 8800 kg honey; 9752 ducklings; 21960 flower and vegetable seedling; 125 q paddy; 54 q vermicompost etc. The products are highly popular among the consumers.
2.	KVK, Nalanda	Mushroom	Mid day meal scheme	KVK Nalanda started promoting mushroom production since 2007 by organizing training programmes and other extension programmes. A few leading producers were also assisted by the district administration to establish spawn production units. This has resulted in increase in mushroom production and drop in price. The growers approached the administration and with the intervention of District Collector mushroom got included in mid day meal scheme, running in schools. This provided the growers a steady market for disposal of their products. As many as 272 SHGs were formed by KVK for mushroom production. Subsequently with the support of KVK and tourism department a '*mushroom production circuit*' was developed involving Nalanda, Rajgir and Bodhgaya districts for better marketing.
3.	KVK, Gumla	Bamboo craft and value added products	Private dealers	A substantial amount of nonfarm forest based resources are available in Gumla. Tribals collect these and sell in local market individually. KVK imparted training in production of value added products *e.g.*, Jamun Powder, Jamun Sirka, Amla Supari, Satawar Jam, Tomato Soup, Honey, Pickles, Guava Jelly, Pear Jam etc. Besides training was also given to women SHGs in bamboo craft and preparation of bamboo based decorative items. The products are being sold by Vikash Bharti, the host organizations from its premises. The KVK has also linked a few private dealers for marketing of these products.

Contd...

Table 11.1–Contd...

Sl.No.	*Name of KVK*	*Crop/ Enterprises*	*Linkage Developed*	*Remarks*
4.	Seva Bharati KVK,Kapgari, West Midnapore	Seed, planting material	Satellite centre of KVK	Agro Service Centre is a well established unit of the SBKVK. It sells seeds, planting materials, bio-products, animal products. It also sells farm implements. Products are highly popular among farming community as well as other Stake holders of district and nearby adjoining State. The KVK has a satellite centre *Sardhanjali Nursery*, which deals with Grafted mango, Guayava, Cashew nut, Patilime, Mosumbi, Litchi, Pomegrante, Ber, Papaya, Cocunut, Banana. Skill Development Training on agri-business and marketing, Regular Monitoring follow-up and guidance is provided by KVK. Annual turn over of this centre is over Rs. 80 Lakhs. KVK has played key role in establishing two more seed ventures *viz.*, *Sabuj Udyan* and *Basundhara* whose turn over are Rs 50 Lakhs and Rs 3 crore respectively. These market linkages have enabled the seed growers realize better price for paddy, mustard and other seeds.
5.	KVK, Hazaribag	Processed food, pickle, phenyl, mushroom etc.	CRPF; Federation	KVK intervened in marketing of products (sauce, pickle, candle, agarbati etc.) produced by women SHGs to Central Reserve Police Force (CRFP) 22 battalion, Hazaribag. Beside holy Cross KVK also runs an outlet for selling various products *e.g.*, washing powder, pickle, phenyl, candle, seeds, planting materials etc.KVK facilitated formation of federations of mushroom grower with over200 members. The federation sell through restaurant, hotel, bakery, vegetable vendors etc. to ensure regular supply and to sustain interest of the consumers the federation at times outsource mushroom from neighbouring districts like Giridih, Ramgarh and Koderma.
6.	KVK, Burdwan	Jute products	District Administration	KVK Burdwan imparts training to rural women for preparation of jute based souvenir items-bag; mobile case; dolls; wall hangings etc. It tagged a few NGOs like *Meghdhoot Welfare Society, Panagarh; Ujjiban, Barddhaman* etc. and Farmers Clubs like *Sonar Bangla Farmers Club; Pragati Farmers Club* etc. for marketing. KVK with the support of NABARD and District Administration established six shops/rural marts at Durgapur, Burdwan and Galsi town. These marts provided a better market outlet for jute products and efforts are on to establish few more.

Contd...

Table 11.1–Contd...

Sl.No.	Name of KVK	Crop/ Enterprises	Linkage Developed	Remarks
7.	KVK, Uttar Dinajpur	Low cost food for children	District Administration; ICDS	Incorporation of the low cost nutritious food developed by Uttar Dinajpur Krishi Vigyan Kendra in a project called "PUSHTI" – An initiative for providing additional nutritional supplements to the malnourished pre-school children through SHG networks of the district. KVK also acts as a resource centre for imparting hands on training to the aspiring SHG members of each ICDS Blocks in preparation of low cost food and sensitizing them in storing of the food as per specification. Linking SHGs with ICDS has provided steady market for the low cost food for children.
8.	KVK, Dakshin Dinajpur	Mushroom processing	Product development and branding	KVK Dakshin Dinajpur trained Sri Chotaru Barman in mushroom growing. He sells mushroom and processed products like – pickle, biscuit, sauce, pakora under the brand 'Hilli Dilli'. District Industrial Center and KVK urged him expand his business by procuring mushroom produced by small scale producers and selling with his brand. As many as 30 producers sell their products to Barman who in turn markets it as dry mushroom and in other variants. Now Hilli Dilli mushroom products are available in distant market like Siliguri, Gangtok, and Malda.
9.	KVK, Godda	Quality seed	Kisan club	KVK has formed several Kisan Club and SHGs in different villages like Lalpur, Diara (Godda block), Pipra, Chilra (Pahargamma block) and Maheshpur (Basantrai block) and with the help of NABARD and the groups are engaged in seed production of paddy, pigeonpea, gram, mustard and elephant foot yam. The foundation seeds of the improved varieties are provided by the KVK. The seeds are registered in the Jharkhand Seed Certification Agency and the Seed Certification Officer visits the farm during the cropping season. The seeds are processed in the processing unit of KVK and sealed in 30 kg bag with proper stitching and tag. The availability of seeds in season are informed to the farmers through newspaper, during training programmes in different villages and through mail to the different government and non-government organizations of the district and outside the district. During the year 2014 - 15, 225 q of paddy, 2 q of gram, 1 q of pigeonpea and 3 q of mustard seeds were sold.

Contd...

Table 11.1–Contd...

Sl.No.	*Name of KVK*	*Crop/ Enterprises*	*Linkage Developed*	*Remarks*
10.	RAKVK, Nimpith	Ornamental birds	Doorstep marketing	KVK organised training for the landless womenfolk of island villages and introduced ornamental bird rearing technology in riverine blocks of Sundarbans, the "*Aila*" affected zone. It also provided diagnostic help and advisory services to the farm women and also regular monitoring of the rearing units. Initially, the island producers sold *budgerigar* and other pet birds at local markets at main land. This way of marketing proved difficult and they failed to realize good price. KVK intervened and facilitated a doorstep marketing channel by which the women can sell their product form their home itself to young entrepreneurs, trained in scientific transport and proper care during carrying. The entrepreneurs collect ornamental birds and sell it in the main land to traders as well as to hobbyists. This link has not only provided easy disposal by the rearers but also job opportunities to a few rural youth.
11.	KVK, Palamu	Lac	Linking traders with farmers	KVK, Palamu has taken initiative in reviving lac cultivation at Dulsulma and Murma village in Satwarba block since last three years. Lac cultivation practices have spread to 12 adjoining villages. There was no market available for disposing lac. With KVK's intervention traders from distant places started coming to weekly market, which was primarily a vegetable market. The traders buy brood lac and stick lac from the weekly market and transport it to other places. On peak harvest season 10-12 qt of brood lac is sold in the market. Linking lac cultivators with traders has proven to be quite useful in promoting lac cultivation in Palamu.
12.	RAKVK, Nimpith	Processed meat	West Bengal Livestock Development Corporation	Goat and sheep farming is very popular to landless rural people because it has advantage of minimum capital investment. Besides, Bengal goats and Garole sheep are highly prolific, have high disease resistance and above all the meat is very popular. However, the farmers do not get right price owing to the presence of too many intermediaries. KVK intervened and provided farmers direct access to West Bengal Livestock Development Corporation, the nodal agency for marketing meat under brands like '*Goat meat'; 'Goat liver'; Lamb meat'* etc. Additionally KVK also provides the facility for minimal processing of the meat as the Corporation does not procure live animals. Farmers are coming to KVK for processing of meat on a specified day and selling the same to Corporation at a pre-determined price. This has improved price realization by farmers to a great extent. Thus creation of marketing linkage given a boost to promoting different animal husbandry activities among the farmers.

Besides, in the field of marketing of farm produces there are several such novel approaches initiated by various stakeholders. Krishi Vigyan Kendras, working in the district level are the frontline extension system providing need based technical support to farming community. Few KVKs have also initiated new approaches of bringing buyer and seller together. Some such cases holds immense potential for replication. The catalytic role played by KVKs in marketing farm produces is worth documenting. The out-of-box solutions developed at the grassroot level needs upscaling by forging forward linkages. A few selected institutional innovations in marketing are mentioned in Table 11.1

Market linkages provided by the KVKs are unique in its nature and function. Both conventional (private dealers, sales counter etc.) and non-conventional (CRPF, ICDS etc.) buyers were tapped by the KVKs. Innovative ways of marketing (doorstep marketing, mushroom circuit etc.) were also initiated. These institutional innovations, piloted with facilitation by KVKs, for marketing of farm produces are new and it would be too early to evaluate their performances.

Satellite centre of KVK established by West Midnapore KVK for propagation of seeds and planting materials is doing commendably well. Facilities for production are being managed by private entrepreneur whereas KVK provide technical support and guidance. This arrangement has solved the problem of seed scarcity to a great extent.

Conclusion

- ☆ There should be paradigm shift in production. Farmer should produce only what market needs. Research fraternity has to acknowledge the fact that farmers' sole attention is on more profit and all out effort in needed to make farming an attractive vocation for the youth. Farming needs to be seen as business and business plans for both farm and non-farm based enterprises will have to be developed for ready consumption by potential entrepreneurs.
- ☆ Market orientation of agricultural production is of paramount importance. Farmers need to be provided with market intelligence, quality standards, export potentials etc.
- ☆ Extension need to embrace ICT in a big way to cater to the information needs of farming community
- ☆ In order to upscale marketing innovations strong linkage will have to be built between producer organizations and marketers.

References

Ameur C. 1994. Agricultural Extension: A Step Beyond the Next Step. World Bank Technical Paper No. 247.World Bank, Washington, DC.

Bebbington H, Thiele G, Davies P, Prayer M and Riveros H. 1993.Non-Governmental Organizations and the State in Latin America.Rethinking Roles in Sustainable Agricultural Development, Routledge, USA.

Chambers R, Pacey A and Thrupp L A. (Eds.) 1989. Farmer First: Farmer Innovation and Agricultural Research. Intermediate Technology Publications, London.

Hall A. 2009. Challenges to strengthening agricultural innovation systems: where do we go from here? *In:* Scoones I and Thomson J (Eds.) Farmer First Revisited: Innovation for Agricultural Research and Development. Practical Action Publishing, Sussex, UK, pp. 30–38.

Hall A, Mytelka L and Oyeyinka B. 2004. Innovation Systems: What's Involved for Agricultural Research Policy and Practice? ILAC Brief 2.

Hall A, Sulaiman V R, Clark N and Yoganand B. 2003. From measuring impact to learning institutional lessons: an innovation systems perspective on improving the management of international agricultural research. *Agric. Syst.* 78 (2): 213–241.

Kaimowitz D, Snyder M and Engel P. 1990. A conceptual framework for studying the links between agricultural research and technology transfer in developing countries.*In:* Kaimowitz D (Ed.) Making the Link: Agricultural Research and Technology Transfer in Developing Countries. Westview Press Inc., Colorado, USA and London, UK, pp. 227–269.

Ortiz O. 2006. Evolution of agricultural extension and information dissemination in Peru: an historical perspective focusing on potato-related pest control. *Agric. Hum. Values* 23: 477–489.

Oscar Ortiz, Ricardo Orrego, Willy Pradel, Peter Gildemacher, Renee Castillo, Ronal Otiniano,Julio Gabrie, Juan Vallejo, Omar Torres, GemebredinWoldegiorgis, BelewDamene, Roger Kakuhenzire, Imelda Kasahija andIgnatiousKahiu 2013.*Agric. Syst.* 114: 73-83.

Rogers M. 1962. Diffusion of Innovations, first ed. The Free Press, New York.

Rogers M. 1995. Diffusion of Innovations, fourth ed. The Free Press, New York.

Umali D, Schwartz L. 1994. Public and Private Agricultural Extension: Beyond Traditional Frontiers. Washington, DC. World Bank Discussion Paper 236.World Bank.

Chapter 12
Gender Inequality in India

Aparna Jaiswal and Arvind Saxena

Department of Extension Education,
JNKVV-College of Agriculture, Ganj-Basoda, Vidisha, M.P.

Introduction

Gender equality and empowerment of women always recognized globally as an important goal in the development agenda. Empowerment is a multidimensional approach. It operates in the social, economical and political spheres of life and at individual, household, community, market and institutional levels.

Gender refers to the social roles and relations assigned between women and men and it includes the different responsibilities of women and men in a given culture or location unlike the sex of men or women which is biologically determined. Gender roles of women and men are socially constructed and such roles can change overtime and vary according to geographical location. Gender norms dictate the role of women and men in different spheres. In some societies it restricts women mobility and engagement in productive work outside their homes and checks their growth. Such norms vary from place to place and are often slow to change. Women and men must be included at all stages any planned action, legislation, making women's as well as men's concerns and experiences integral dimensions in the design, implementation monitoring evaluation of all agricultural policies and programmes so that women and men benefit equally. It is very essential to reduce the gender disparity for reducing women's drudgery and for bringing gender equity. Women and men should have equal conditions for realizing their full rights and potential to contribute to national, political, economic social and cultural development and benefit from results. The *World Development Report 2012: Gender Equality and Development* (World Bank, 2012c) argues that greater gender equality is not just the right thing to do but also smart economics. It also highlights the need to improve the availability of quality gender-disaggregated data and supports more experimentation and systematic evaluation.

India's economic growth has been impressive from the last few decades but gender indicators and human development provide continuing cause for alarm. India's progress on gender equality lags behind progress in comparison to other areas in terms of education, health, economic development, and violence against women. Though more women are going out to work, India primarily remains a nation of stay-at-home wives who play a major role in keeping families together in a country with virtually no government-aided social security. With so many million women out of the workforce, India's overall ranking reflects unbalanced government policies that are failing to harness the full potential of a key demographic.

Keeping the above points in mind this paper an endeavor has been made to understand the demographic changes and the other aspects of gender inequality from available data on education, health and employment.So that it will be helpful for decision makers in amending the existing policies.

Methodology

The present study is based on the secondary data obtained from official source of report and survey statistics of Census of India 2011, Women and Men in India, 2011 and articles on Census of India 2011, demography statistics and figures of different Census of India.

The Gender Equality Index

The Gender Equality Index (GEI) is a new index for measurement of gender disparity. This index is a composite measure which captures the loss of achievement within a country, due to gender inequality, and uses three dimensions to do so: reproductive health, empowerment, and labor market participation. The GEI capture the loss in achievement due to gender disparities in the areas of reproductive health, empowerment and labour force participation with values ranging from 0 (perfect equality) to 1 (total inequality).

Global Gender Gap

Global Gender Gap focuses not on empowerment of women, but on the relative gap between men and women in four fundamental categories - economic participation, educational attainment, health and survival, and political empowerment. It includes measures such as estimated sex selective abortion, number of years the nation had a female head of state, female to male literacy rate, estimated income ratio of female to male in the nation, and several other relative gender statistic measures. It does not include factors such as crime rates against women versus men, domestic violence, honor killings or such factors. Where data is unavailable or difficult to collect, World Economic Forum uses old data or makes a best estimate to calculate the nation's Global Gap Index (GGI).

Social Institutions Gender Index

Social Institutions Gender Index is a measure of discriminatory social institutions that are drivers of inequalities, rather than the unequal outcomes themselves.

Gender Statistics

The following table compares the population wide data for the two genders on various inequality statistical measures, according to The World Bank's Gender Statistics database for 2012.

Gender Statistic Measure	*Females (India)*	*Males (India)*	*Females (World)*	*Males (World)*
Infant mortality rate, (per 1,000 live births)	44.3	43.5	32.6	37
Life expectancy at birth, (years)	68	64.5	72.9	68.7
Expected years of schooling	11.3	11.8	11.7	12.0
Primary school completion rate, (per cent)	96.6	96.3		
Lower secondary school completion rate, (per cent)	76.0	77.9	70.2	70.5
Secondary school education, pupils (per cent)	46	54	47.6	52.4
Ratio to males in primary and secondary education (per cent)	0.98	1.0	0.97	1.0
Secondary school education, gender of teachers (per cent)	41.1	58.9	51.9	48.1
Account at a formal financial institution, (per cent of each gender, age 15+)	26.5	43.7	46.6	54.5
Deposits in a typical month, (per cent with an account, age 15+)	11.2	13.4	13.0	12.8
Withdrawals in a typical month, (per cent with an account, age 15+)	18.6	12.7	15.5	12.8
Loan from a financial institution in the past year, (per cent age 15+)	6.7	8.6	8.1	10.0
Outstanding loan from banks for health or emergencies, (per cent age 15+)	12.6	15.7	10.3	11.6
Outstanding loan from banks to purchase a home, (per cent age 15+)	2.26	2.35	6.6	7.4
Outstanding loan from banks to purchase a home, (per cent age 15+)	2.26	2.35	6.6	7.4
Unemployment, (per cent of labor force, ILO method)	4	3.1		
Unemployment, youth (per cent of labor force ages 15-24, ILO method)	10.6	9.4	15.1	13.0
Ratio to male youth unemployment rate (per cent ages 15-24, ILO method)	1.13	1.0	1.14	1.0
Employees in agriculture, (per cent of total labor)	59.8	43		
Employees in industry, (per cent of total labor)	20.7	26		
Self-employed, (per cent employed)	85.5	80.6		
Cause of death, by non-communicable diseases, ages 15-34, (per cent)	32.3	33.0	29.5	27.5
Life expectancy at age 60, (years)	18.0	15.9		

Table 12.1: Population of India 2011 at Glance

Population	*Statistics*
Total Population	1,21,01,93,422 (persons)
Males	62,37,24,248
Females	58,64,69,174
Decadal Growth (2001-2011)	18,14,55,986 (17.64 percent)
Sex Ratio	940 Females/1000 Males
Density of Population	382 per sq. km.
Literacy (in per cent)	Total; 74.04, Males: 82.14, Females: 65.46

Source: 2011 Census Data.

As per Census 2011, India is the second populous country of the world after China which accounts for world's 17.5per cent population. The population of India is 1210.19 million comprising 586.47 million (48.5per cent) females and 623.72 million (51.5per cent) males. Females have a share of 48.1 per cent in the urban population and of 48.6per cent in the rural population. The average annual exponential growth rate of population during 20012011 is 1.64per cent. The decadal growth rate during the same period stands at 17.64per cent which is 3.90per cent less than that during 19912011. The 2001-2011 periods is the first decade (exception of 1911-1921), which has actually added lesser population compared to the previous decade. The sexratio (number of women per 1000 men) is 940 in 2011 which shows continued improvement over the sex ratios of 927 in 1991 and 933 in 2001.

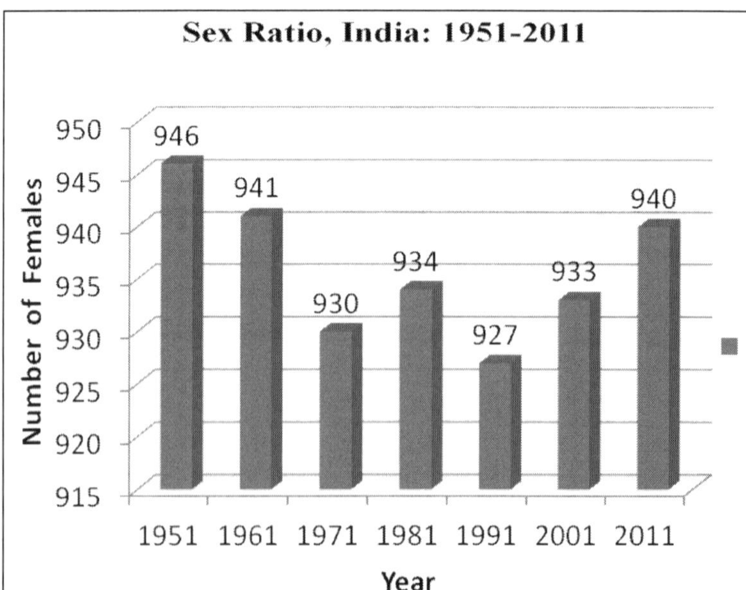

Key Finding-This wide gap in sex ratio clearly indicates the strong preference of male children over female. Another darkside of the finding is legislations attempt to ban sex determination and female foeticide are not effective.

Table 12.2: Global Rankings: India's Global Rank on various Gender Inequality Indices

Indicator	*India's Global Rank (Year)*	*Source*
GII (Gender Inequality Index	132/148 (2012)	United Nationals Development Programme (UNDP)
GGI (Gender Gap Index)	101/136 (2013)	World Economic Forum (WEF)
SIGI (Social Institutions and Gender Index	56/86 (2012)	Organisation for Economic Co-operation and Development (OECD)

UNDP has published gender inequality index, India ranks 132 out of a total of 168 countries in 2012.

According to the Global Gender Gap Report released by the World Economic Forum (WEF) in 2011, India was ranked 113 on the Gender Gap Index (GGI) among 135 countries polled. India has improved its rankings on the World Economic Forum's Gender Gap Index (GGI) to 105/136 in 2013 (Global Gender Gap Report 2013). When broken down into components of the GGI, India performs well on political empowerment, but is scored to be as bad as China on sex selective abortion. India also scores poorly on overall female to male literacy and health rankings. India with a 2013 ranking of 101 had an overall score of 0.6551, while Iceland, the nation that topped the list, had an overall score of 0.8731 (no gender gap would yield a score of 1.0).

Alternate measures include OECD's Social Institutions Gender Index (SIGI), which ranked India at 56th out of 86 in 2012, which was an improvement from its 2009 rank of 96th out of 102.

Key Finding-India has failed badly in removing gender-based disparities, ranking so poor in World Economic Forum's 2014 gender gap index, scoring below average on parameters like economic participation, educational attainment and health and survival.

Review

2005 UN study found that social norms-based gender discrimination leads to gender inequality in India.

Babu and Babu (2011) concluded that In India, dowry is the payment in cash or some kind of gifts given to bridegroom's family along with the bride. The practice is widespread across geographic region, class and religions.

Begum *et.al* 2007 found in Madurai, India that old age security, economic motivation, and to a lesser extent religious obligations, continuation of the family name, and help in business or farm, were key reasons for son preference. In turn, emotional support and old age security were main reasons for daughter preference. The study underscored a strong belief that a daughter is a liability.

Gender Gap in Education: Missing Ingredients of Inclusive Development

For social and economic development education is the most valuable factor. It is crucial to promote education among all backward sections those have fallen behind the national average in the field of education for the up gradation of society. Education not only opens up opportunities for individual and productive employment but also makes them capable to participate in various areas of social life. The world community recognizes education for all as an important Millennium Development Goal of which India is signatory. Over the years the gender gap in education has been narrowing, yet substantial disparity still exists and there is an urgent need to be overcome.

According to the released data Census of India 2011, literacy rate has gone up from 64.83 per cent in 2001 to 74.04 per cent in 2011 showing an increase of 9.21 percentage. Literates constitute 74 per cent of the total population aged seven and above and illiterates form 26 percent. Percentage growth in literacy during 2001-2011 is 38.82; males: 31.98 per cent and females: 49.10 per cent. While literacy among males rose from 75.26 per cent to 82.14 per cent, an increase of 6.9 per cent, it rose among females from 53.67 per cent to 65.46 per cent, an increase of 11.8 points. Of the additional literates, women (110,069,001) outnumber men (107,631,940). The gap of 21.59 percentage points between men and women in 2001 now stands reduced to 16.68 per cent.

Table 12.3: Literacy Gap by Sex (Percentage)

Year	*Male*	*Female*	*Total Literacy Rate*	*Male Female Literacy Gap*
1951	27.16	8.86	18.33	18.30
1961	34. 44	12.95	24.02	21.49
1971	39.45	18.70	29.45	20.75
1981	46.89	24.82	36.23	22.07
1991	64.10	39.30	39.3	24.80
2001	75.30	53.70	64.8	21.60
2011	82.14	65.46	74.04	16.68

Source: Census of India, 1951-2011.

Key Finding-Above data clearly indicates that in spite of a boom in the education sector, Indian women also remain less educated than men even though they make up nearly half of the population.

Review

As per the report of "Women and Men in India" (2012) the main reasons of females never attending school are 'expensive cost of education', 'not interested in studies', 'education is not considered necessary' and 'required for household work'.

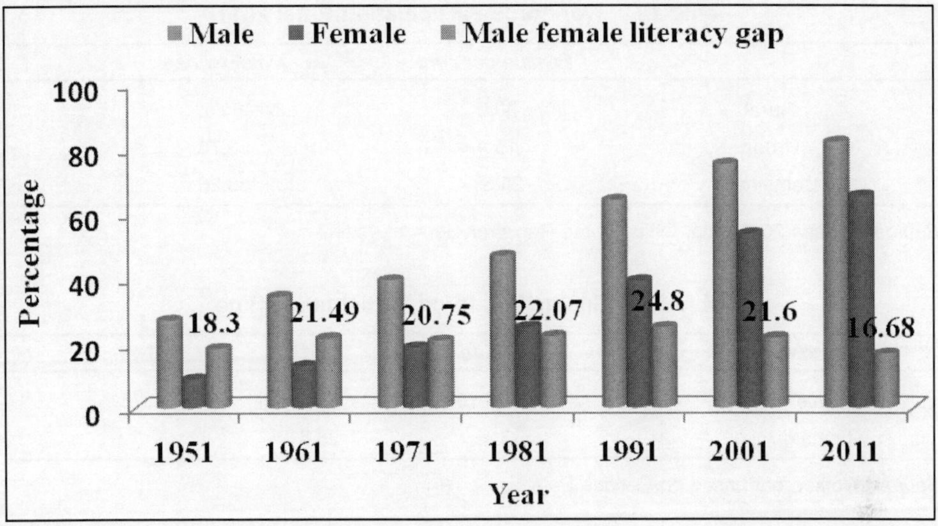

"Human Development Report for 2012" reported that from 2006-2010, the percent of females who completed at least a secondary education was almost half that of men, 26,6 per cent compared to 50.4 per cent.

As per the report of "Wage Rates in Rural India" (2008-09) Men dominate the numbers of those enrolled in higher educational degrees.

Kalyani and Kumar 2011 revealed that however literacy rate is gradually rising, the female literacy rate in India is lower than the male literacy rate.

Workforce Participation Rate

Workforce development system is a way to serve the needs of organizations communities and nations. Workforce development as coordinated policies and programs that collectively "enable individuals the opportunity to realize a sustainable livelihood and organizations to achieve exemplary goals, consistent with the history, culture, and goals of the social context"(Jacobs and Hawley,2003).It is an urgent need to motivate and count women's participation equally in powering the growth of Asia's third largest economy, which can have a multiplier effect in eradicating "Policy makers should encourage women's participation in powering the growth of Asia's third largest economy, which can have a multiplier effect in eradicating poverty and illiteracy and poverty.

As per Census 2011, the workforce participation rate for females is 25.51 per cent against 53.26 per cent for males and the gender gap in workforce participation rate is 27.75 which was 26.10 per cent in 2001. Rural sector has a better female workforce participation rate of 30.02 per cent compared with 53.03 per cent for males whereas for urban sector. The participation rate of females trails at 15.44 per cent against 53.76 per cent for males. 41.1per cent of female main and marginal workers are agricultural labourers, 24.0 per cent are cultivators, 5.7 per cent are household.

Table 12.3: Workforce Participation Rate: 2011

	Female per cent	*Male per cent*
Rural	30.0	53.0
Urban	15.4	53.76
Combined	25.51	53.26

Source: Census 2011 data, Office of the Registrar General, India.

Table 12.4: Gap in Workforce Participation Rate

Gender	*2001*	*Gap*	*2011*	*Gap*
Male	51.7	26.10	53.26	27.75
Female	25.6		25.51	

Source: Worked out based on Census Data.

In the same period, the average wage/salary received by regular wage/salaried employees of economically active age group was Rs. 428.66 per day for females compared with Rs. 550.23 per day for males in rural areas. For urban areas, it was Rs. 609.7 and Rs. 805.52 per day for females and males.

Gender gap in workforce participation rate clearly indicates that women, as half of the human capital of India, will need to be more efficiently integrated into the economy in order to boost India's long term competitive potential also there is a general disparity in wages and earnings between men and women for a similar work. So, women need employment justice.

Review

Wage Rates in Rural India (2008-09) reported wage inequality between men and women in India. The largest wage gap was in manual ploughing operations in 2009, where men were paid Rs 103 per day, while women were paid Rs 55, a wage gap ratio of 1.87. For sowing the wage gap ratio reduced to 1.38 and for weeding 1.18. For other agriculture operations such as winnowing, threshing and transplanting, the men to female wage ratio varied from 1.16 to 1.28. For sweeping, the 2009 wages were statistically same for men and women in all states of India

Lakshmana 2007 found significant gender-gap between the All India wage rates of both agricultural and non-agricultural sector and corresponding rates for Karnataka and M.P.

Rao stated that about 78 per cent of rural women are engaged in agriculture, compared to 63 per cent men.

Gender Sensitive Policies of the Government

The Indian constitution has gender equality and women's rights issues:

- Article 15(3) empowers state to make affirmative discrimination in favour of women

- ✰ Article (39) - directs state policy towards providing men and women equal right to means of livelihood and equal pay for equal work
- ✰ Article (42) directs state to make provisions for ensuring just and humane conditions of work and maternity relief
- ✰ Article (51a) imposes fundamental duty on every citizen to renounce practices derogatory to the dignity of women
- ✰ Tenth Plan (2002-2007) focuses on empowering women socially and economically and eliminates gender discrimination.

Different states and union territories of India along with the cooperation with the central government, have initiated a number of region-specific programs for women empowerment to reduce gender inequality over the 1989-2013 period.

Some of these programs include Swarnajayanti Gram Swarozgar Yojana, Sampoorna Gramin Rozgar Yojana, Awareness Generation Projects for Rural and Poor women, Condensed Course of Education for Adult Women, Kishori Shakti Yojana, Swayamsidha Mahila Mandal Programme,[Rashtriya Mahila Kosh, Support to Training and Employment Programme for Women, Swawalamban Programme, Swashakti Project, Swayamsidha Scheme, Mahila Samakhya Programme, Integrated Child Development Services, Balika Samriddhi Yojana, National Programme of Nutritional Support to Primary Education (to encourage rural girls to attend primary school daily), National Programme for Education of Girls at Elementary Level, Sarva Shiksha Abyhiyan, Delhi Ladli Scheme and Ladli Laxmi Yojana others.

Conclusion

Gender inequality has adverse impact on development goals as it checks economic growth. It hampers the overall welfare because blocking women from participation in social, political and economic activities adversely affect the whole society. Wide sex ratio is a great matter of concern for the nation. It is essential to change the mindset of people simultaneously with legislation strictness for female foeticide. A well educated population is not only essential to support economic growth, but is also a precondition for growth. Separate policy for female education can raise the literacy level to some extent. As a result of discrimination against women, their work is low status, low paid and lacks potential upward mobility. Women carry the heavier work burden in food production, but because of gender discrimination they get lower returns for their work. Gender inequality proofs as a major hindrance in nation's growth and if constant gender inequalities continue, there is a need to again rethink concepts and strategies for bringing equality.

Policy Makers must Formulate Policies in these Aspects to Promote Gender Equity

- ✰ Ensure equal access to basic education and skills development opportunities for girls and young women
- ✰ Support entrepreneurship for women through training, skill development lowering gender barriers to credit, creating women friendly and conducive business environment, strengthening agricultural extension and business services.

- Improve gender equality in access to basic services like schools, health care, child care etc by investing in supply of infrastructure, staff etc, so as to reduce women vulnerability and increase their capacity to access more remunerative jobs and opportunities.
- Recognition of labor work of working women in the rural economy may be accounted in monetary terms.
- Ensure decision making of women in social, economic and political life of the nation.
- Priority must be given to women in accessing credit on soft terms from banks and other financial institutions for business, for buying properties, and for housing etc.
- Taking serious steps to prevent crime against women and ensuring for a safe environment for women.

References

Babu and Babu (2011), Dowry deaths: a neglected public health issue in India, Int. Health, 3 (1): 35-43

Begum and Singh; CH Sekher and Hatti, (2007) Unwanted Daughters: Gender discrimination in modern India7

Dijkstra and Hanmer (2000), Measuring socio-economic gender inequality: Toward an alternative to the UNDP gender-related development index, Feminist Economics, 6(2): 41-75

Gender Gap Report 2011World Economic Forum, page 9

Gender Statistics,The World Bank (2012)

Global Gender Gap Report 2013. World Economic Forum. Retrieved 31 March 2014.

Human Development Report for 2012. United Nations Development Project. Retrieved 31 March 2014.

Jacobs, R. and Hawley, J. (2003). *Workforce development: Definition and relationship with human resource development.* In Lynham, S. and Egan, T. (Eds.) AHRD Conference Proceedings,Bowling Green, OH: Academy of Human Resource Development.

Kalyani Menon-Sen and A. K. Shiva Kumar (2001). *"Women in India: How Free? How Equal?"*. United Nations. Archived from the original on 2006-09-11. Retrieved 24 December 2006.

Lakshmana,C.M(2007). Demographic change and Gender Inequality: A comparative study of Madhya Pradesh and Karnataka.Institute for Social and Economic Change,Working Paper 183.

Periodic Review: India report 2005. United Nations. Retrieved 28 April 2014.

Rao, E. Krishna, (2006), Role of Women in Agriculture: A Micro Level Study. Journal of Global Economy, Vol 2

Social Institutions and Gender Index: India Profile. OECD. Retrieved 31 March 2014.

Tisdell, Roy and Ghose, (2001), A critical note on UNDP's gender inequality indices, Journal of Contemporary Asia, 31(3): 385-399.

Wage Rates in Rural India (2008-09) Labor Bureau, Ministry of Labour and Employment, Govt of India (2010).

Women and Men in India, 2012, 14th Issue.

World Bank. 2012a. *Core Sector Indicators—Guidance. Results Platform. OPCS, 2012.* Author, Washington, D.C

www.worldbank.org

Chapter 13
Women Empowerment through Farm Mechanization

Joginder Singh Malik and Nitu Sindhu

*CCS Haryana Agricultural University,
Hisar – 125 004, Haryana*

The UN (2001) defined women's empowerment in terms of five components: "women's sense of self-worth; their right to have and determine choices; their right to have access to opportunities and resources; their right to have the power to control their own lives, both within and outside the home; and their ability to influence the direction of social change to create a more just social and economic order, nationally and internationally".

The significance of the role women in development had been recognized by the government of India right from the first five year plan (1951-1956). However, women in these earlier plans were considered as subjects of 'welfare' and category of disadvantaged groups such as destitute, disabled, aged etc. With the passage of time, 11th and 12th five year plan have focused on inclusive growth. Despite the various discrimination faced by women, the 12th five year plan has emphasized on single women by providing 'quota' in jobs. Moreover, women's inclusion in inclusive growth is only then possible when inclusive democracy turned to be reality not the myth.

The Empowerment of women has become one of the most important concerns of 21st century not only at national level but also at the international level. Efforts by the Govt. are on to ensure Gender equality but Government initiatives alone would not be sufficient to achieve this goal. Society must take initiative to create a climate in which there is no gender discrimination and Women have full opportunities of Self decision making and participating in the Social, Political and Economic life of the

Country with a sense of equality. Beside this we must not forget to make the women of our country well aware of their rights.

The vision for the XII Five Year Plan is to ensure improving the position and condition of women by addressing structural and institutional barriers as well as strengthening gender mainstreaming.

Goals for the XII Five Year Plan

- ✰ Creating greater 'freedom' and 'choice' for women by generating awareness and creating institutional mechanisms to help women question prevalent "patriarchal" beliefs that are detrimental to their empowerment.
- ✰ Improving health and education indicators for women like maternal mortality, infant mortality, nutrition levels, enrolment and retention in primary, secondary and higher education.
- ✰ Reducing the incidence of violence against women and providing quality care services to the victims.
- ✰ Improving employability of women, work participation rates especially in the organised sector and increased ownership of assets and control over resources.
- ✰ Increasing women's access to public services and programmes through establishing and strengthening convergence mechanisms at multiple levels, creation of physical infrastructure for women and improving the capacity of women's organizations and collectives.
- ✰ Ensuring that the specific concerns of single and disadvantaged women are addressed.

When we talk about women empowerment it is very clear that economically, women have been making progress, but still men enjoy a large share of the cake. Women are discriminated on numerous fronts including education and training that are essential useful weapons in the competitive world of knowledge for productivity, production, equity, sustainability and empowerment.

While looking at the scenario of women empowerment we cannot neglect the most powerful women of our country, Indra Nooyi (CEO of Pepsico), the Congress President Sonia Gandhi, 1994 Miss World Ashwaryia Rai, the famous tennis player Sania Mirza, Medha Pataker founder of Narmada Bachao Aandolan, Arundhati who won the coveted Booker Prize in 1997 for her first novel The God of Small Things, Kiran Bedi who is the first woman to join the Indian Police Service, Mata Amritanandamayi who is a widely respected humanitarian and known as the hugging saint, Irom Chanu Sharmila known for her campaign against the controversial Armed Forces (Special Powers) Act, 1958 in Manipur and many more. When we look at all there empowered women we can't neglect the fact that all these women come from the highly privileged urban society of India. At the same time when we look at the women in rural India, we find them struggling for the basic requirements for their survival. Things like water, food, education, medical facilities, employment and status is still lacking. When one section of our society is well aware of their facilities and rights, and other section is not even fully aware about them how can we expect to become a nation where women are empowered. There is huge gap between the rural and urban section of our society. This gap has to be reduced, for this creating awareness will play major role.

Rural women and girls have many roles and many responsibilities. They are farmers, care-givers, wage labourers and micro-entrepreneurs and they often spend many hours fetching water and collecting firewood. The empowerment of women is fundamental to reduce poverty, hunger and malnutrition. Gender equality and women's empowerment are important factors for the social and economic development of a nation. The promotion of gender equality and empowering of women is one of the eight Millennium Development Goals (MDG) to which India is a signatory.

As more women are left to undertake the labour-intensive farming sector, the country is in direct needs to provide lighter, efficient, economical and user-friendly equipment to the female population. In rural areas the male migrate oversees for jobs, the women are the one who is working on the farm but due to the lack of access to proper agricultural machinery ranging from tillage, ploughing, plantation, harvesting, processing and storage to women farmers, labour costs have soured, subsequently decreasing production. Compared to male counterparts, the females have equally contributed to the agricultural sector in the country. But mass migration among the young population has added to the burden of the female population over the years. Poor access to mechanical tools and technologies used for the agriculture sector, however, has made life even more difficult for women farmers. The increasing work burden on women is further aggravated by unpredictable weather and rainfall phenomenon.

A majority of the women farmers are still dependent on more traditional and manual forms of agricultural tools that are labour-intensive, time-consuming, heavy and inefficient. The use of iron and wooden ploughs, disc harrow, spade, hoe and sickle are still common among them. Majority of women farmers across the country rely on subsistence farming and have limited financial capacity to own and operate available machines.

A report made public by the United Nations' Food and Agriculture Organization (FAO) in January 2014 said that the farm machines have gradually revolutionized agriculture and reduced drudgery for the millions of farming families across the globe. Krishna Veni and Seehtalakshmi are two farm women of Azhivaikkal village near Thanjavur. They can now drive a tractor with multi-crop seeder fitted to it and sow seeds in fields thanks to the training given by International Rice Research Institute (IRRI). They demonstrated their skill at the International Women's Day organized by the IRRI, jointly with Thanjavur Janaseva Bhavan, a voluntary organization which has credit linked hundreds of women's self-help groups in the area at a field at Azhivaikkal village near Thanjavur on Wednesday. They drove the tractor in front of a large number of farm women and farmers. R. Ganeshmoorthy, Assistant Scientist, Extension Agronomy, Cereal Systems Initiative for South Asia (CSISA), IRRI, India, explained the technology to farm women. Mr. Mc Guinness said that IRRI had recognized the role of women farmers in food security and livelihood issues. More than 50 per cent of the agricultural operations like transplanting and harvest were carried out by women.

Another case is from Bangladesh where farmers went from using human muscle and ox power in the early 1970s to being one of the most mechanized agricultural economies in South Asia, with 300,000 low-power two-wheel tractors and a million diesel powered irrigation pumps. We need to come up with special programmes focusing on the need of female farmers and make farm implements available to them. Until they acknowledge the vital role of women in agriculture and provide them with opportunities in the form of efficient and economical machinery to address the problem, the improvement of overall agricultural sector remains slim. Meanwhile, efforts are being made to promote research, development and promotion of mechanizing the sector to address the growing concerns arising with the drop in production and increase in fallow land over the years.

Empowerment of women can give power to women to have control over the circumstances of their lives, control over resources and ideology. The resources could be physical, human, economic, intellectual, social, political and/or religious resources. The control over ideology means ability to influence decisions, thinking and situations which is reflected in values, attitudes and beliefs. Empowerment includes greater self confidence and an inner transformation of one's consciousness that enables one to overcome external offices. Empowerment of women is critical not only for their own welfare but also for the development of the country.

Indian envisions a future in which Indian women are independent and self reliant. In various national policies and developmental programmes, emphasis has been given to empower the women in all spheres of life. United Nations Fund for Women in its guidelines on women's empowerment interpreted empowerment as a process where women, individually and collectively become aware of how power relations operate in their lives and gain the self-confidence and strength to challenge gender inequalities at the household, community, national, regional and international levels.

In previous plans, there was not much more attention on women's cause. But it is widely believed that men enjoy all the power, position and privileges women were treated as second rate and lower human spheres. The theory of patriarchy puts the question of women in a proper perspective. Low income single women are vulnerable and victims of the patriarchal structures that deprive them of inheritance and other rights. The Census of 2011 estimates that 833 million people continue to live in rural India and a very large proportion of them are either wholly or significantly dependent on crop agriculture, horticulture, animal husbandry or fisheries. The Mahila Kisan Sahakatikaran Pariyojana (MKSP) which is a sub-component of NRLM (National Rural Health Mission) was recently launched to meet the specific attention needs to be given to leverage the enormous potential of empowering rural women in the farm sector, and also in the rural sector, and also in the rural non-farmer sector.

Thus, we can say that women's discrimination is phenomenon in Indian society. In fact, still now they are ill treated by their family in terms of low intake of food, lack of educational attainments and exploitation both in public and private spheres. We know that despite the half of the population, their voice was not counted in various decisions making bodies. Democracy is based on the principle of equality. So, bringing Indian women in the mainstream is need of the hour. Provision of "quota" for single women in jobs and some other positive discrimination for women may create a gender-just society in near future. But there are other issues which should be considered which are making women aware of their rights. Awareness programmes should be started for women in rural areas so that women empowerment reaches the core of our society where it is needed the most.

References

Women Empowerment and Challenges in India. Ramesh Shantaram Mhatre, International Journal of Entrepreneurship and Business Environment Perspectives Pezzottaite Journals, Volume 2, Number 1, January-March' 2013

Report of the stearning committee on women's agency and child rights for the twelth five year plan (2012-2017) planning commission government of India.

Faster, Sustainable and More Inclusive Growth, An Approach to the Twelfth Five Year Plan (2012-17), planning commission government of India.

The Paradox of Gender Responsive Budgeting, Bhumika Jhamb, Yamini Mishra, Navanita Sinha. may 18, 2013 vol.38, EPW Economic and Political Weekly

Socio-economic empowerment of women through 'Science, Technology and Innovation Policy, 2013' – the missing link?* Sunita Sangar CURRENT SCIENCE, VOL. 106, NO. 8, 25 APRIL 2014

Gender Issues for Technological Empowerment of Women in Agriculture, DARE/ ICAR ANNUAL REPORT 2006–2007

Gender gaps and Women's Empowermentin India–Issues and Strategies Amtul Waris and B. C. Viraktamath International Journal of Scientific and Research Publications, Volume 3, Issue 9, September 2013 1 ISSN 2250-3153.

Women in twelfth five year plan in India: An Analysis Satarupa Pal A journal of Humanities and Social Science Published by: Dept. of Bengali Karimganj College, Karimganj, Assam, India Website: www.thecho.in

Chapter 14

Newer Approaches for Technology Transfer in Livestock Sector with Special Reference to ICT Tools

Rupasi Tiwari and Triveni Dutt

ICAR-Indian Veterinary Research Institute, Izatnagar, Bareilly – 243 122, U.P.

The livestock sector has emerged as one of the key components of agricultural growth in developing countries. Policy makers in India today are finally acknowledging the shift in the agriculture sector. Economic contribution of livestock is today more than that of food grain crops. In fact in the year 2002-03, monetary contribution of livestock for the first time surpassed that of food grains, and has since remained higher by 5-13 per cent. In fact, both livestock and fisheries components have been growing faster than the crops component for a decade. The report for the 12th Five Year Plan has also recognized this shift. Livestock now provides a quarter of the agriculture gross domestic product (GDP). In 2010-11, it generated output worth Rs 3,40,500 crore (at current prices). This was 28 per cent of the agriculture GDP and about 5 per cent of the country's GDP. The total output from livestock was higher than the value of food grains (Rs 3,15,600 crore) and fruits and vegetables (Rs 2,08,800 crore). After livestock, paddy is the next highest contributor to the agriculture GDP. In 2009-10, output from livestock was 2.5 times the value of paddy and more than thrice the value of wheat, as per the Central Statistical Office data. In fact it is said that animals act as an insurance against income shocks of crop failure and natural calamities. Small and marginal farmers, landless labourers and women are more dependent on livestock for supplementing incomes and generating gainful employment in rural areas.

Policy makers are taking a serious note of this new economy. It is now seen as a major support for the crops sector to project decent overall agriculture growth. As per the note of the Economic Advisory Council on fiscal outlook for 2010-11, it estimated an optimistic agriculture growth based on the growth in the livestock economy. The report stated that despite the drought of 2009, growth in the livestock sector will ensure that the agriculture sector growth will be stabilized. Further the 12th plan document report of the working group on animal husbandry and dairy also reported that "The livestock sector is expected to emerge as an engine of agriculture growth in the 12th Plan and beyond in view of rapid growth in demand for animal food products". (Mahapatra,R, 2012)

Egalitarian and Food Security Function of Livestock

Further it has been found that rural poverty is less in states where livestock contributes more to farm income. Punjab, Haryana, Jammu and Kashmir, Himachal Pradesh, Kerala, Gujarat and Rajasthan are a case in point (Planning Commission report, 2012-2017). Mostly, marginal farmers and those who have quit farming are joining the livestock business. About 70 per cent of the livestock market in India is owned by 67 per cent of the small and marginal farmers and by the landless. In fact it can be said that prosperity is now more dependent on per capita livestock ownership than on agricultural farms.

In terms of food security, the significance of livestock and dairy products becomes enormous. Livestock rearing plays a critical role in the socio-economic development and welfare of India's rural population not only as a source of food (milk, meat and egg) but also draught power, manure, hides, skin, hairs, bones, blood, fat, drugs etc (Sharma and Tiwari, 2011).

Livestock Resources and Production

India's livestock resource is one of the largest in the world. It has 512.05 million livestock, which includes 190.904 million cattle, 108.702 million buffaloes, 135.17 million goats 65.06 million sheep, 10.29 million pigs, 0.62 million horses and ponies, 0.19 million mules and 0.32 million donkeys, 0.4 million camels, 0.07 million yaks, 0.29 million mithun beside 729.2 million poultry (19th Livestock census, 2012).India ranks first with respect to buffalo, second in cattle and goats, third in sheep and fifth in poultry population in comparison to the world livestock and poultry population. Total Livestock population in India has increased from 292.8 million in 1951 to 530 million in 2007 at the rate of 1.06 per cent. In contrast, population of horses and ponies, camels, pigs, mules and donkeys has shown a trend of negative growth. Further the country produced 137 million tones of milk in 2013-2014 with annual growth rate of 5 per cent. The total egg production also reached to 74.75 billion eggs in 2013-2014 from 63.02 billion in 2010-2011. The poultry meat production is estimated to be about 2.69 million metrictonnes. The current per capita availability of eggs is around 61 eggs per year.

Exports of poultry products are currently at around Rs. 457.82 crore in 2011-12 as per the report of Agricultural and Processed Food Products Export Development Authority (APEDA).The Wool production declined marginally at the end of Eleventh Five Year Plan (2011-12) to 44.7 million kg from 45.1 million kg in the Tenth Five Year Plan (2006-07). The Annual growth rate for production of wool is about 4 per cent in 2011-12 compared to previous year. The meat production has registered a healthy growth from 2.3 million tonnes at the end of Tenth Five Year Plan (2006-07) to 5.5 million tonnes at the end of the Eleventh Five Year Plan (2011-12). The Annual growth rate for meat production in 2011-12 was about 13 per cent. Total export earnings from livestock, poultry and related products were Rs.33,417 crore during 2011-12 as compared to Rs.25,409 crore during previous year *i.e.* 2010-11 with a growth of about 31.5 per cent. Total export earnings from fisheries products have also shown steady increase and reached to Rs. 16,597.23 crore during 2011-12 compared to Rs.12,901.47 crore during 2010-11 with a growth of about 28.65 per cent.(Annual Report, DAHD and F, MOA, 2012-2013 and 13-14)

Problems in Indian Livestock Industry

Global food demand is ever increasing and expected to be doubled by the year 2050 due to exponential increase in population while the agricultural productivity is declining, largely due to shrinking of resources particularly cultivable land. In this situation, role of livestock sector to fulfill growing food demand is of paramount importance in terms of nutritional security and employment generation. Low productivity, health and production threatening diseases, weak infrastructure of Indian dairy and animal husbandry sector, marketing constrains, adulteration etc pose major challenge to the Indian Livestock sector. Due to the low productivity of cattle, the cost of milk in India is high as compared to the International market (Hegde, N.G. 2001). The mean milk yield of Exotic/Crossbred Cows is 6.78 kg/day, while that of Indigenous/Non-Descript Cows is 2.50 Kg/day and Buffalo 4.91 kg/day, and of goats Goat 0.45 kg/day (Annual Report 2014-2015, DAHD and F,MOA, GOI). Cattle milk yield is about half of the world average and about 12 per cent to 15 per cent of that in the USA, Canada and Israel. (Birthal and Rao, 2002). Greater efforts are needed to improve per animal productivity through emphasizing propermanagement efforts. Besides management, animal health plays a major role in the progress of livestock industry. Occurrence of diseases causes heavy economic losses resulting due to mortality, less efficient resource use and need of faster stock replacement rates, all of which requires additional financial as well as personnel resources. Control and treatment of diseases also add to the losses. India alone losses 4.45 billion US dollars (Rs.20, 000 crores) per year due to the uneconomic performance of the animals because of adverse effect of some diseases. Unless sincere efforts are made, prevention and control of common diseases like Foot and mouth disease, mastitis, haemorrhagicsepticaemia, brucellosis, tuberculosis, paratuberculosis, IBR, leptospirosis, anthrax, black quarter, tetanus, rabies, pseudorabies, PPR, swine fever etc. of animals, the country is far away from achieving the optimum targeted production of 160-170 million tones of milk by 2020. Various technologies have been developed in the livestock sector to enhance the productivity of the livestock and improve their health. These technologies have remarkably reduced the morbidity and mortality rates and enhanced the production of our livestock. Still lot needs to be done to reduce the huge economic losses incurred by farmers due to the direct and indirect effect of animal diseases. Further, feed and fodder shortage in the country is another major challenge facing the livestock sector. Other challenges include low productivity, weak infrastructure of Indian dairy and animal husbandry sector, marketing constrains, adulteration etc

Approaches in Livestock Extension

Great research efforts in the country have been able to develop new livestock technology for raising the production level of farmers but it is the strategy of technology transfer of the nation that determines the translation of scientific achievement to practice at farmers' level. Increased priority to technology-induced growth in the livestock subsector would improve food and nutritional security, alleviate poverty, and reduce interregional and interpersonal economic inequities. Since independence, a number of strategies have been formulated and implemented for transferring the technical knowhow to the livestock owners.

The planned extension efforts in livestock and dairy sectors started with the launching of community development programme and formation of National Extension Service in 1950's. The community development programme launched in the first five year plan emphasized on increasing the milk production along with enhancing food grain production in the country. The other livestock development programmes during this period were launching of cattle breeding policy, key village scheme and public milk supply scheme. With the launching of second five year plan the cattle breeding policy was more clearly defined and it included specified selective breeding in case of well-defined breeds and the use of male progeny for the development of non-descript cattle. The Intensive Cattle Development Programme was launched in 1964-65 and it covered all the areas of livestock development *viz.* breeding, feeding and management of cattle. The major emphasis was on producing high yielding breeds of cattle. The launching of AICRP in 1969 on various livestock species was a major effort in the area of livestock development. But the breakthrough in livestock development occurred with the launching of Operation Flood-I.

The other extension approaches for livestock development during the 1970's were the special livestock development programme such as small farmers' development agency (SFDA), marginal farmers and agricultural laborers (MFAL) and various area specific programmes such as the drought prone area programme (DPAP), Hill Area Development Programme (HADP) etc. The integrated rural development programme launched in 1979-80 gave importance to popularizing the livestock rearing among the resource poor by providing assistance in form of loan, training and input supply. The other extension programme that have led to the development of livestock sector are the technology mission on dairy development, the ICAR's transfer of technology projects *viz.* the Lab to Land Programme (LLP), Operational Research Project (ORP), KrishiVigyan Kendra (KVK) and the IVLP (Institute Village Linkage Programme) (Tiwari, R. 2005)

Newer Extension Approaches in Livestock Sector

Continuous researches in the area of animal health and production in the country have been able to develop new livestock technologies for raising the production level, but not all of these technologies have been effectively transferred to end users. Numerous extension approaches have been evolved and tested for transferring the agricultural technologies to the field level, since pre-independence. Although many of them are still effective, but newer approaches need to be developed and tested in the changed social scenario with latest techniques and tools for faster and wider dissemination of technologies in lesser time with appropriate feedback. Our newer approaches need to focus more on enhancing per animal productivity, reducing the disease incidence and helping in marketing of livestock produce so as to make the sector more remunerative. Some of the approaches already in use in certain parts of India and abroad and some new approaches have been described here that would help improve the production and productivity of the livestock sector.

Extension Approach for Reducing the Economic Losses Due to Diseases

Introduction of exotic breeds in India have increased the incidence of diseases especially Foot and Mouth Disease (FMD), Haemorrhagic Septicaemia (HS), *peste-des-petits* ruminants (PPR), Brucellosis, Mastitis, Blood protozoon diseases etc. India losses 4.45 billion US dollars (Rs.20, 000 crores) per year due to the adverse effect of performance of the animals because of some diseases (*The Cattle Site News Desk*). Reducing the incidence of occurrences of these diseases can be achieved by conducting intense campaigns. Campaign is a coordinated communication and educational effort focusing attention on a particular problem and its solution over a period of time. Campaigns have been successful in the past in bringing about noticeable changes in development. Further, campaigns have been found to be very effective in producing behaviour changes and attitudinal changes among the people. Such intense activity can help in eradication of epidemic diseases affecting livestock. Government of India has implemented many such intense campaign programs on controlling livestock diseases which pose a major threat for animal husbandry economy. Assistance for States in Controlling Animal Diseases, National Rinderpest Eradication Program (NREP), Foot and Mouth Disease Control Programme, Professional Efficiency Development (for field vets) etc., are some of the important programs implemented by the government. (www.dhad.nic.in). NREP was a model program for intense campaign against Rinderpest. It is imperative to say here that India was declared "Rinderpest free" in March 1998, which was possible only because of this intense campaign. Similar programs should be implemented in reducing the occurrences of dreadful, economy threatening diseases like FMD, Brucellosis, PPR etc by means of installing awareness hoardings in all the villages, regular advertisements in radio, television, exhibitions, special field visits by the specialists teams of Veterinary institutions etc. Pulse polio vaccination program is one such successful program still in progress. Similar programs for livestock diseases are to be initiated. (Tiwari, *et al.*, 2012)

Extension Approaches for Conservation of Indigenous Technologies and Native Breeds

The folk wisdom towards livestock rearing and maintenance are getting extinct in the near future. Techniques, technologies and methods in treating livestock which happened to be a style of life in the past are not seen now in many places. Inappropriate aid for conservation, changes in agriculture, technology, economy and disasters are said to be the possible reasons for the extinct of native breeds (www.love4cow.com, 2011). Introduction of new livestock cross breeding policies at central and state level administrations has led to the destruction of native breeds. Cross breeding has although gifted increase in milk production however it has reduced disease resistance among the livestock population.

Tradition based livestock treatments are liable to be the first choice for many farming communities (Ghotge*et al.*,2002). The reach ability of modern veterinary care is said to be only 20 per cent in spite of the efforts taken by government and non - government veterinary health agencies. Hence a need for conserving the native techniques and breeds is gaining paramount importance among the policy makers and many projects have been formulated and are in progress. Many NGOs like Anthra, BAIF and National Innovation Foundation are working in this direction of conserving tradition based native livestock technologies.

a) Three Tier Model: An Experience from KVK, Ahmednagar

A Three-tier model of technology assessment, refinement and dissemination has been followed by KVK Ahmednagar wherein with the involvement of farmer clubs, the KVK conducts field assessment and refinement of the technologies experimented by the farmers. Thus refined technologies are disseminated and promoted to all the 150 villages where the KVK has interference. These activities are carried out with the help of 561 Self help groups, who are attending regular training programs in the KVK. The Ahmednagar KVK has obtained a positive response in technology dissemination by this particular three tier model (Gaikwad, 2011).

b) Incentives for Rearing Native Breeds

Native breeds of cattle are said to be having good adaptability for the Indian climatic conditions and better resistance over many infectious and contagious diseases. Cross breeding has although gifted increase in milk production however it has reduced disease resistance among the livestock population. Milk production ability of Gir, Sahiwal, Red Sindhi, Tharparker are comparatively better than other local ND breeds of India. Hence, as a measure to conserve native germ plasm and breeds, Government should take initiatives to create awareness among the people about the importance and specialties of native germ plasm and encourage the farmers to rear native breeds. Since the production parameter aspects of native breeds will not stand at par with the cross bred animals, the farmers may neglect to rear native breeds. Providing proper incentives for those who rear native breeds will help the conservation of these indigenous breeds of cattle.

Digital/ICT Based Extension Approaches for Man Power and Investment Problems in the Present Scenario

Present day facilities for animal health care are inadequate and needed to be much improved. At present, there is one veterinary center for every 14299 animals which is too low in comparing the recommendation of National Commission on Agriculture (NCA) to have one Veterinary Center for every 5000 animals, and therefore, most of the states are lagging far behind the NCA recommendation. Accordingly, it is necessary to increase and strengthen the existing veterinary health care facilities to take care of precious livestock. There are only 8,732 veterinary hospitals/polyclinics and 18,830 veterinary dispensaries in the country providing service for the huge livestock population. Limited availability of man power and weak extension system has increased the work burden on the veterinary officers. Further research results of the countrywide tele-consultancy network for farmers "Kisan Call Centre" level-II node at the Indian Veterinary Research Institute, catering to animal husbandry related queries for two stated of UP and UKD indicate that the livestock owners lack information with regards to scientific health care of the livestock which is very necessary to gain an optimum production from the livestock. And therefore more concentration of efforts are required to educate the livestock owners for the health care of the livestock on priority and more information packages need to be developed for them. Especially queries on mastitis were highest among the infectious disease which reveal its occurrence and severity among the dairy owners. Among the reproduction related queries most important queries were on problem of repeat breeding followed by anoestrus and prolapse of uterus for which information package needs to be prepared and provided to livestock owners. Among the management/production/nutrition related aspects majority of the calls were on low production of cattle and buffaloes. The analysis of the queries received at the Kisan Call Centre at various other nodes in the country related to livestock production can form a major base for gathering and prioritizing the information needs of the livestock owners and for developing need based information packages for them. (Tiwari *et al.*, 2010). To address this problem, Information Communication Technology mediated tools and cyber extension has emerged and is being used widely to reach more people in short period of time. (Tiwari *et al.*, 2012)

Digital/ICT Tools for Livestock Extension

At present there is an IT revolution and the introduction of computer has changed the whole style of teaching and learning. Even the way the information is to reach to the farmers has drastically changed. The time taken to send the information has reduced and the quality of information transmitted has improved commendably. There are various ways in which the ICT tools are helping the farmers. Examples are the use of SMS for agro advisory, web based advisory/discussion forums, e-mails, CD based information packages etc. One important characteristic of these media is that the learner can learn at his own speed and these require less number of extension workers. In fact we can group these methods under Cyber extension. Cyber extension is an information exchange mechanism over cyber space, the imaginary space behind the interconnected computer networks through telecommunication means. It utilizes the power of networks, computer communications and interactive multimedia to facilitate information sharing mechanism (Wijekoon, 2003). The ICT tools/digitized information systems are gaining more and more importance now a day, as they create interest in the users thereby resulting in effective communication and transfer of technologies. Some of the ICT enabled tools being used and certain newer tools which can be made use of for the livestock extension are mentioned below:

i) Expert System

Computer-based system designed to respond like a human expert in a given field. Expert systems are built on knowledge gathered from human experts, analogous to a database but containing rules that may be applied to solving a specific problem. An interface allows the user to specify symptoms and to clarify a problem by responding to questions posed by the system. Software tools exist to help designers build a special-purpose expert system with minimal effort. An outgrowth of work in artificial intelligence, expert systems show promise for an ever-widening range of applications. There are now widely used expert systems in the fields of medicine, personnel screening, and education. Various expert systems are available for farmers and livestock owners such as Poultry expert system, Buffalo reproduction expert system, Expert system for coconut pests etc. BRIS is a knowledge base information retrieval system for buffalo reproduction developed by IVRI, Izatnagar. This system covers different aspects of buffalo reproduction, which includes information on male and female reproductive system as broad subjects and deal individually. This system can assist the field veterinarians in understanding the buffalo reproduction system. 'Poultry Expert System' (PES) was developed on selected dimensions of poultry farming. The Veterinarians and Veterinary students found the user attributes of PES of greater utility, less complexity, moderate compatibility good technicality, feasibility and user friendly (Thammi Raju and Rao, 2006). 'MakeFeed', is windows based software developed by the scientist of Central Avian Research Institute for formulation of balanced feed for wide variety of poultry birds like layer, and broiler chickens, quails, guinea fowls and dairy animals etc. for maximum production performance. (Nath. *et al.*, 2002). Another feed formulation software Pig feed formulator (PFF) has been developed for assisting the commercial pig farmers for ration formulation PFF is powered to handle 25 feed ingredients including mineral mixture, common salt and vitamin mix which are commonly available in the most parts of the country. Additionally the software package has tutorial video with the audio backup to help the beginners showing, "how to formulate balanced feed." The Programme is developed in two languages, English and Hindi. (Tiwari, R., 2013)

ii) Digital Information Packages for Livestock Development

These are systems that provide information to users in a multimedia format with combination of audio-visual, text, animation, graphics and pictures. The interactivity of the system makes it more attractive and user friendly. Many such systems have been developed at IVRI for livestock owners. These include "Livestock and Poultry disease information System" (LPDIS) in English and "Pashudhanavumkukkut Rog SuchnaPranali" (PAKRSP) in Hindi has been developed which focuses on the 78 major diseases of livestock (Cattle, Buffalo, Sheep, Goat, Pig, Horse) and poultry and nine common package of practices for sustainable livestock and poultry production. Both the systems are a comprehensive package containing information about all the infectious, non infectious, metabolic, parasitic, and fungal diseases of major livestock species and Poultry. The major feature of this information system is its highly appealing presentation along with audio backup which helps in inculcating greater understanding of the diseases among the end users. This system is specifically designed to give information on each disease under four major heads *i.e.*, Epidemiology, Symptoms, Treatment, Prevention and control, to the livestock owners and other people engaged in the treatment and health care of livestock as well as to the students engaged in degree or diploma courses in animal health disciplines. (Hindustan Times, Oct., 23, 2011, New Delhi). Two more interactive softwares one for the Para Veterinarians and stockman entitled "Animal Health Information System" (AHIS) in english and another for the farmers of Maharashtra entitled "Health information system for dairy animals" in Marathi has been developed at IVRI, Izatnagar for animal health management (Phand*et al*, 2009, Tiwari *et al.*, 2010). Development of a compact Disc entitled "Pashuchikitsa and PashupalanPrashnottri" at the IVRI, Izatnagar is yet another effort to provide solutions to farmers on their frequently asked queries through e- module, which is easy to search and operate.A software for providing the farmers information about important reproductive problems found in buffaloes in the Northern Gangetic plain agro-climatic region was developed at IVRI entitled "BhainsPrajananSoochnaPranali (Hindi)". Its a user-friendly, audio visual and stand alone interactive information system.Various other such information system software and expert system could be developed based on the needs of different clients and locale (Tiwari, R. 2013)

iii) Use of Touch Screen Kiosk at Village or Institutions

Ramkumar *et al.* (2003) reported that, demand-led information deliveries through the Touch Screen Information kiosk installed in Veterinary institution of Pondicherry helped farmers in preventing diseases and improving health of animals, saving time, labour and money for this disadvantaged group. They further reported change in adoption practices included steps in clean milk production especially washing of udder with dilute potassium permanganate solution and washing of hands by the milker after obtaining information through Touch Screen Information kiosk at Pondicherry.

iv) ICT Enabled Farmers Help Centers

VijayBabu *et al.* (2011) in their study on perception of dairy farmers on using ICT tools to avail dairy information, have reported that there was a significant difference of knowledge levels on feeding, breeding, management, herd health and related problems on dairy; between the ICT and non-ICT user respondents who had frequent contacts with the officers at farmers help centers. This study was conducted in Puducherry Union territory. This reveals that, the ICT enabled advisory system is effective and the farmers use to have a constant contact with the farmer help centers. Hence, establishing more help centers with enough ICT enabled information tools for farmers shall enable the knowledge enrichment of livestock and agricultural famers.

v) Milk Marketing through ICT Usage

"Warana Wired Village" project of Maharashtra are using existing computer network for milk marketing of dairy cooperatives. The co-operative societies in the village have been ICT enabled to increase the farmers' capacities for information exchange in a government-driven experiment to detect the potential of rural ICT. The funding for this ICT enabling unit is arrived at three sources *viz.*, National Informatics Centre (50 per cent), Government of Maharashtra (40 per cent) and Warana cooperative societies (10 per cent). Using this platform, around seventy villages nearby to Warana are sharing information regarding crop marketing data, new government programs, education opportunities etc. (FAO, 2005).

vi) Agro-advisory Websites and Portals

Multitude of efforts has been launched to provide farmers and livestock owners' information about agriculture and related field through websites and portals. These portal provide multiple functions as well such as they provide the needed information, even have question answer section, even allow farmer to post his query or his experiences and also sometimes the answers are sent to his mobile. The Department of Agriculture and Cooperation (DAC) has over the years undertaken several IT initiatives such as AGMARKNET, SEEDNET, DACNET etc. Similarly, States/UTs have also either developed or are in the process of developing IT applications such as AGRISNET under various programmes of DAC. All such applications will be integrated with the Central Agri Portal (CAP) and the State Agri Portals (SAPs) envisaged under NeGP-A. The Central Agri Portal (CAP) and State Agri Portals (SAP) will also have the options of providing online feedbacks by the stakeholders. This would not only improve transparency but at the same time would help in efficient monitoring. The already existing web portal such as AGMARKNET, DACNET, SEEDNET, and DAC work in the following way. AGMARKNET Provides information on (i) prices and arrivals of agricultural commodities at various mandis in eight Indian languages; (ii)trend analysis of prices and arrivals at market/district/state levels; and (iii)grading, packaging, standards, sanitary and phyto sanitary requirements, and marketing charges. Over 300 commodities and 2000 varieties from 1500 markets are being covered. In all, 3700 markets were proposed to be covered during the Eleventh Plan. The DACNET portal aims to provide information and services to the farming community on a number of specific subjects which include oilseeds, pesticides, Separate portals of Crops Directorates, Pesticides Registration Information System, Plant Quarantine Information System, Website concerning farm machinery provides information on machinery tested, training schedules and modules and application forms for training, Website on organic farming and Frequently Asked Questions (FAQs) on bio-fertilizers/organic farming. DAC Portal (www.agricoop.nic.in) portal provides information on schemes and programmes of the Department, Acts, Rules, reports, publications, documents, conferences, seminars, crop condition, reservoir levels, weather forecast, minimum support prices, etc. The SEEDNET India portal aims to bring all seeds users under a closed-user-group and establish 'Seed informatics online' and develop database on availability, supply and demand for different areas facilitating the optimum utilisation of seeds in the country.

AGRISNET; The main objective of Agrisnet project is to create a sustainable data bank of all the Agricultural elements and agricultural inputs of agricultural sector like climate, rainfall, soil, seed, plant, crop, land, fertilizer, pesticide, market information etc. in one place in every state. Since the main motto of Agriculture department is to increase the productivity and overall development of farmer community in the state that is why Agrisnet project was envisaged to effectively use the Information Technology for dissemination of agriculture related information to the farming community.Eg.AGRISNET of Himachal Pradesh, Arunachal Pradesh, West Bengal etc.

Apart from these government initiatives various other initiatives of private organizations are serving the farmers.TCS' Mobile Agro Advisory System (mKrishi) connects farmers with an ecosystem that empowers them to make sound decisions about agriculture, drive profits and conserve the environment. Farmers require information on weather, soil, fertilizer and pesticide that are specific to their plot of land. They need information on the type of seeds, crops available in the market and local market prices.Kisan School is yet another portal to disseminates fresh agriculture information to 200000 small, medium and large sized farmers, both in a timely and personalized manner through training, voice and text message (SMS, peer-to-peer (P2P) and interactive voice response system (IVRS) using mobile and internet technology enabling farmers to make wise decision on how to produce, what to produce, when to produce and for whom to produce. Kisan School also generates more than 10000 employments for rural youths. (www. kisanschool.org) There are 80 Web sites and portals run by Governments providing information to farmers on soil, plant protection, diseases, weather forecast, fertiliser use and farm and crop management. The number could be much more in the private sector. All this is likely to confuse the farmer. To tackle this, the Union Government has come up with the idea of a unified portal, to provide information on all farming needs. This is the Farmers Portal www.farmers.gov.in launched by DAC and FW, MOA and FW, GOI. The portal would help farmers access all the relevant content related to his district/ village/block. The information is made available in form of SMS, email, audio and video and in the required language.

vii) Use of Short Message Service (SMS)

Next to the Radio and Television the Mobile phone users are increasing rapidly in India particularly in rural areas, which is creating platform for information dissemination through value added services like Short Message Service (SMS). At the end of 2008 there were 347 million mobile subscribers of which rural India contributed around 92 million i.e 26 per cent. The overall teledensity as of end of December 2008 was 33.23 per cent. The mobile phone looks like tomorrow's most likely access device for information dissemination. Already various companies like Thompsons Reuters Market Light (RML) have initiated such service in the rural areas to cater to agriculture community. RML currently operates in nine Indian states with a subscription base of over 175,000 (by the end of 2009). RML SMS cover localized weather forecasts, crop advisory, proximate market data and crop prices; in addition to relevant policy and national and international news (www.globalhand.org). Various KVK in the country have also started their SMS in the form of either text or voice messages.

To reduce the farmers' confusion and to have a common platform for all agro-advisory service providers the Government of India has launched a countrywide mKisan SMS Portal on July 16, 2013. It enables all central and State government organizations in agriculture and allied sectors to give information/services/ advisories to farmers by SMS. The farmers can register to this service by calling Kisaan Call Center on the toll free number 1800-180-1551 or through the web portal and ask for SMS in regional language. The portal was launched with a target of 120 million farmers initially across the country and has till now sent out 1.16 billion advisories. The portal, which has around 7 million registered farmers can send customized messages in 12 regional languages by harnessing the expertise of 3,000 officials and experts of central and state government department. Experts from 130 agricultural universities and government organizations are registered on the portal (Srivastava and Mehta, 2014)

viii) Replicating Kissan Kerala Model

One another ICT based initiative in this direction is "Kissan Kerala". This is a mobile based agricultural information services support for the farmers by the Government of Kerala. To tap the potential use of high mobile network in Kerala and to avoid the language barriers, this SMS based service was established. Information regarding plantation, livestock farms, weather, and market prices of agricultural commodities etc., can be obtained by sending SMS to 9496113311. On subscribing, Kissan Kerala will provide recorded voice based information to the mobile phone users (www.kissankerala.net/)

ix) Use of Discussion/Internet Forums

An Internet forum, or message board, is an online discussion site where people can hold conversations in the form of posted messages. A discussion forum is hierarchical or tree-like in structure: a forum can contain a number of sub forums, each of which may have several topics. Depending on the forum's settings, users can be anonymous or have to register with the forum and then subsequently log in order to post messages. On most forums, users do not have to log in to read existing messages. The discussion forums can help farmers and livestock owners by providing them a platform to post their queries that are answered by the experts.

x) CAI for Livestock Owners

The teaching through CAI (Computer Assisted Instructions) is basically in one of two ways: either it provides a straightforward presentation of data or it fills a tutorial role in which the student is tested on comprehension. Sometimes it does both. The CAI can be effectively used to teach farmers and livestock owners, since it is mostly the adults who differ significantly with each other in their learning speed and ability and further they are shy and feel embarrassed easily if they give an incorrect answer in front of others. Now a day many of the websites provide this type of learning experience and have facility for evaluation also (Tiwari, R., 2013).

xi) Use of Social Networking Sites

Most social network services are web-based and provide means for users to interact over the Internet, such as e-mail and instant messaging. These allow users to share ideas, pictures, posts, activities, events, and interests with people in their network. These also help the farmers to be in touch with the experts and expert groups thus enabling higher and more up to date exchange of information.

Extension Approaches for Cost Recovery for Livestock Services

Livestock services could either be a public good or a private good. Veterinary services displayed varying degrees of public and private good characteristics. Among the livestock services, clinical diagnosis and breeding are examples of private goods, whereas services such as disease surveillance, quarantine and food hygiene/ inspection are public goods. Veterinary services and provision of veterinary inputs formed the health services, while animal breeding, livestock research and extension are production services (Kathiravan*et al.*, 2009). The total amount sanctioned in the union budget 2011 – 12 for the department of animal husbandry dairying and fisheries projects was 160000 crores (www.dahd.nic.in). The cost incurred by the government in providing livestock services are increasing every year. To reduce this, the authorities are taking measures to recover the cost from the clientele.

i) Pubic Private Partnership in Animal Husbandry

Encouraging private corporate companies to provide livestock services would be an alternative measure in reducing the cost incurred by the government. Public Private Partnership means an arrangement between a government/statutory entity/ government owned entity on one side and a private sector entity on the other, for the provision of public assets and/or public services, through investments being made and/or management being undertaken by the private sector entity, for a specified period of time, where there is well defined allocation of risk between the private sector and the public entity and the private entity receives performance linked payments that conform to specified and pre-determined performance standards, measurable by the public entity or its representative.

PPP has been initiated in many parts of the country to develop dairy and animal husbandry activities. PPPs can serve as a major step towards development of people and economy of India on the whole. The PPP models have been overwhelmingly successful in driving a major wave of urban infrastructure development but now the onus lies on the Government to expand the scope of these partnerships into new horizons of rural development. Public private partnerships have been initiated in major thrust areas of dairy industry like procurement, infrastructure and logistics development, operations management, capacity building through training and extension research and knowledge transfer (Chand *et al.*, 2010).

Government of Uttarakhand has started a special wing for pubic private partnership. With the help of this special wing on PPP, the government has initiated many animal husbandry development projects. Government of Rajasthan has committed to improve its fiscal by prioritizing improvements in thrust areas including animal husbandry through public private modes. Operations were proposed to be initiated to improve the productivity status of livestock and agriculture for which the actions would be taken in PPP fashion. Governments of Madhya Pradesh and Andhra

Pradesh have started to encourage public private partnership models in improving the states' food processing industries.

ii) Government–Farmer Cost Sharing Efforts

Cost-sharing, which involves government-farmer partnership in the funding of agricultural extension service, is one of the reforms aimed at achieving sustainable funding for extension systems. Ozor *et al.* (2007) reported that in Nigeria majority of farmers (80.6 per cent) and extension professionals (85.7 per cent) had favourable perceptions towards cost-sharing. In India also some Non Governmental Organizations such as Bharatiya Agro Industries Foundation, J.K Trust, GayatriShikshasadanSamsthan etc., have come into existence by providing some veterinary services on cost recovery basis and are found to be more efficient than the Government machinery (Satish and Prem Kumar 1993). Further, cost recovery efforts by various veterinary universities, and cooperative services have shown the capacity of farmers to pay. A study by Ravikumar *et al.* (2010) in Andhra Pradesh revealed that large holders identified more veterinary services as suitable for cost recovery than the small and medium holders. They suggested that the services which provide direct individual benefits *viz.*, production and curative services could be brought under cost recovery, and, services which are of public service type such as prophylactic and extension services could be provided free by Government. More of such studies in other states would help in developing proper cost recovery measures for the livestock services.

Extension Approaches for Marketing of Livestock Products

As per the WTO norms and the formation of global market, the competition in marketing livestock products has considerably increased among the nations. Produce of sub-standards lose their hold in the arena. Apart from the production aspects, handling of the produce gains momentum in the recent era. According to reports loss in the animal husbandry sector due to poor handling of the products including storage, transportation, value addition etc. So, it is imperative to know that high profit can be earned by increasing the value of the raw produce through further processing and preparation of products in a proper manner. The requirement here is to change the present strategies and emphasizing on value addition of the livestock produce and develop markets for these produce.

i) Market Led Extension

In agricultural and animal husbandry context, the demand for the products and its market size is getting constantly expanded but the marketing system is not regularized in that pace. Direct marketing in animal husbandry and agricultural produces, *i.e.*, producer to consumer marketing is negligible. Loss due to poor marketing chain is reported to the tune of Rs 50,000 crores/year. There are huge problems encountered in storing, grading, standardizing and processing the produces. To overcome all these hindrances, market led extension is gaining importance in the recent past. Market led extension is the market ward orientation of agriculture and animal husbandry production through extension education. Regular market analysis, market oriented production, updating market knowledge and intelligence are the important aspects of market led extension. The focus of the extension functionaries need to be extended beyond production. Farmers should be sensitized on various aspects on quality, consumer's preference, market intelligence, processing and value addition and other marketing information. This will help the farming community to realize high returns for the produce, minimize the production costs, and improve the product value and marketability. (Tiwari *et al.*, 2012)

ii) Establishment of Certified Farmers/Dairy Owners Market

A Certified Farmers' Market is an approved site where farmers/Dairy owners will sell their produce directly to consumers. These markets and the participating farmers should be approved by the Government. The California state of the US has California Farmers Market Association and in various counties of the state, there is Certified Farmers Market. In these markets the growers are prohibited from buying and re-selling produce. In order to sell at a Certified Farmers' Market, a grower must obtain a Certified Producer's Certificate from their home county's agricultural commissioner's office. This certificate must be displayed at the point of sale and must list each commodity offered for sale. The specific County Division of Agriculture routinely inspects farmers' markets to ensure growers are selling only the items they are authorized to sell. (http://www.sccgov.org). Establishing such markets for dairy and other products at various places in the blocks can ensure that producers get good remunerative price for their produce and the consumer gets a fair product. Such markets will also help in preventing malpractices such as milk adulteration, since these markets will at least help in organized sale of milk to the consumer, thereby offering a suitable platform to conduct inspection of the products traded in the market (Tiwari *et al.*, 2012)

iii) The Rural Business Network: A Network Model in United Kingdom

The Rural Business Network is a joint venture between ADAS, the National Farmers Union (NFU) and the Country Landowner's Association (CLA). The aim is to provide a communication and information network for all rural industries and interests in the UK and beyond. The Rural Business Network (RBN) is a separate company, owned and funded by ADAS, the NFU and the CLA. The system is hosted on a commercial agricultural network service called 'Farming on Line' (FOL) based in Leicestershire, central England. FOL provide the servers and local call access throughout the UK and are responsible for the development, installation and support of the client software on user's PCs. The client software enables the user to download material for later use off-line and the majority of the client code is also held locally. The arrangement with FOL has several advantages including, providing a high quality client environment and access to a range of features and services which would have been time consuming and costly for the RBN to develop in-house.

Marketing data are provided in spreadsheet format (the RBN contains a spreadsheet emulator to obviate the need for users to have spreadsheet software on their machines) to enable them to be downloaded and analyzed locally. There are also simple but extensive email facilities both within RBN and via the Internet. The technology has proved to be remarkably resilient and the stability of the dial-up links has been very good - even on lines to very remote areas. The off-line facilities are certainly a valuable selling point although the attraction seems to be currently to email and communication rather than extensive viewing of reference material. Significant interest is being generated by farmer and grower groups who will use the system as a communication tool initially but who may require additional services once they become familiar with the technology (Andy L Offer, 1997).

Extension Approaches for Motivating the Extension Personnel

A technology transfer program will be a successful one only if the desired technology is fully accepted and adopted by the target clientele. The adoption rate of scientifically proven, economically viable animal husbandry technologies are said to be meager. According to reports, adoption rate of artificial insemination in cattle is about 35 per cent (Innovations for Poverty Actions Project, 2011). Very successful, scientifically proven, recommended all over the India, the artificial insemination technique's adoption rate is very low and this shows the situation of other technologies developed. To overcome this problem, apart from fulfilling the prerequisites of the characters of an innovation, motivation of end users holds immense significance. Motivation is a process in which the desire to do things are stimulated internally and or externally. It's the crucial element in setting and attaining goals—and research shows you can influence own levels of motivation and self-control. Few methods have been discussed here regarding motivation aspects.

i) Introducing Inter-village and Intra-village Competitions

Introducing competitions between villages such as the clean village, the ectoparasite free village, the disease free village, the calf mortality free village, the best livestock health village etc. will lead to better efforts by the gram pradhans and other panchayat members to strive for success of the village. Further, such competition within villages among the livestock owner such as ectoparasite free farm/disease free herd/clean and hygienic shed may also lead to more awareness and initiation among these farmers to excel in such practices leading to higher productivity.

ii) Village Adoption Programs

Adopting a village by a group and providing special attention towards the growth and development of the village will result in better diffusion of the new innovation and technologies for the farming community. Many leading veterinary institutions like IVRI, TANUVAS etc have adopted villages and helped the farmers to better adopt the newer technologies in animal husbandry. The very idea behind the village adoption program would be the focused attention of the university activities in the villages which would be a major motivating factor for the beneficiaries.

Extension Approaches for Doorstep Services

In order to provide timely inputs and other services to the farmers at their door step mobile institutions were initiated. Since the services are timely available in the doorsteps of the beneficiaries, these models have proven to be success. Still to enhance the effectiveness of these services few innovations in their methodology is proposed.

i) GIS Enabled Mobile Veterinary Clinics

Mobile Veterinary Units are the service sections manned by Veterinary Officers providing veterinary health services namely, treatment, vaccination, de-worming and breeding coverage to livestock reared by farmers residing in remote areas who face difficulties to reach veterinary institutions. These Units go around their area of operation on a scheduled programme. The services provided by these Units are well recognized by the public as they provide benefits at the doorsteps of the farmers. This facility is available with many state government departments including Tamil Nadu (55 units), Andhra Pradesh (15 units), Karnataka (174 units) etc. Apart from the government sector, private practitioners do service at the farmers' proximity.

These services are often time scheduled and does not cater the inevitable emergencies. Hence, a GIS enabled service delivery should be structured and utilized for the benefit of the farmers. In this model, enough number of mobile veterinary units traceable by the district hub through GIS mode should be wandering in the service area. Call from the farmer would reach the district hub, where the officials shall identify the farmer's location and nearby serving mobile unit through GIS. Both should be concurrently intimated, and the service would be provided to the needy at limited time frame. This system would cater the need based service, and the initial investment cost could be taken back by the implementing agencies by charging fees for the service. (Tiwari *et al.*, 2012)

ii) Mobile ATIC Vans

In order to motivate and create awareness about the newer technologies to the far flung resident farmers, Agricultural Technology Information Center, IVRI, Izatnagar, has launched 'Mobile ATIC van facility' on 22nd September 2010. These mobile ATICs are a miniature form of the main ATIC at the Izatnagar campus in all the aspects. These mobile ATIC vans contain exhibits of latest technologies in the field of animal husbandry. Pamphlets, leaflets, inputs etc are provided to the clients. People in need of further information regarding livestock technologies are advised to visit the main campus. The very basic theme behind this "ATIC on Wheels" is that the clientele are to be motivated to have an access to the new technologies developed and promoted by the institution at their door step (www.ivri.nic.in).

iii) Satellite ATICs

In the service of farming community, at present there are around 47 ATICs (ATIC interface, 2010) across the nation. These 47 ATICs are sufficiently providing information for those who do visit the institutions. In order to extend the services provided by the ATICs, some ten to fifteen satellite ATICs can be established in the rural institutions like panchayat offices, educational campuses, PDS outlets, community halls, farm of progressive farmers etc. These satellite ATICs under the charge of trained professionals can work like miniature ATICs guided by main campuses. By this way a wide network of single window information system services delivery can be provided to a larger mass of clientele (Tiwari *et al.*, 2012)

Extension Approaches through Rural Institutions

Rural Institutions play critical role in translating agricultural and livestock production into profit. In effect, these institutions could be a potential agent of change in the rural areas. The neglect of these institutions resulted into deprivation of access to information, technology, other means of economic gain, and increased the probability of their exclusion from agriculture. Several models of modern institutions such as Panchayati Raj Institution (PRI), Rural Retail Chains, Post offices, Community centers and the Self-help groups (SHG) are being experimented in various parts of the nation (Policy paper 39, NCAP, 2007). Some approaches to make use of these rural institutions in effective dissemination of technology are discussed below.

i) Networking with Post-offices for Technology Dissemination

Indian Agricultural Research Institute has initiated an approach to utilize the vast connectivity of post offices for its technology dissemination. Seed distribution through post offices was carried out which fetched a good response from the needy farmers. By this mode, seeds of bottle gourd, paddy, pigeon pea, bajra and mustard seeds were disseminated across the 20 adopted villages under five selected post offices in Sitapur district of Uttarpradesh. (Vijayragavan, K., 2011). Utilizing this vast network of post offices in various activities of livestock development can be initiated by providing subsidized mineral mixtures, deworming drugs etc., for the farmers.

ii) Transfer of Technology Clubs

The technology transfer clubs were initiated by NABARD in rural areas for transmitting latest agricultural techniques to the farmers at their field level. This approach orients them to establish better relationships with the funding agencies, adoption of latest technologies, value addition etc. and ensure the benefits of collective bargaining in procurement and marketing of inputs and produce respectively. Sree*et al.*(2011) in their study on the status of technology transfer clubs in west Godavari district of Andhra Pradesh have reported that there were 181 clubs in the district and those clubs are the grass root level informal forums of farmers which were organized by various stake holders like rural banks, NGOs research stations and KVKs. The research institutions arrange regular training programs on awareness of newer technologies, projects on mini dairies, tribal development activities, sericulture, orchard, organic farming etc. which has enhanced the interest and involvement of the members towards knowledge generation on newer techniques in agriculture and allied activities.

Other Approaches

Pluralistic Extension Approach

A key concept in current discussions of various aspects of rural development is pluralism, the realization that farmers are best served by the broadest possible array of information sources and structures. The complexity of rural development demands a plurality of technological solutions and service structures, including a readiness to continuously unpack packages as situations change and become better understood. The purpose of focusing on pluralism is to take into consideration the jumble of ongoing activities and, rather than trying to gain control over them, to instead choose niches and to identify common concerns where different approaches may lead to synergy. Awareness of pluralism allows extension planners to admit that they cannot co-ordinate all the variables and perform all the functions. Acknowledging pluralism in implementing agencies and approaches may focus attention on the platform approach as an alternative conceptual model for adaptive planning and allocation of scarce public sector extension resources. It is a way of moving beyond making wish-lists of all the things we would like the extension service to do (Ian Christoplos, 1996).

Area and Species Specific Development Approach

The government of Rajasthan has initiated programme for development of various models of livestock production systems, improved feeding practices through managerial interventions for effective utilization of feed and fodder, integrate traditional and modern approaches and technologies for each district for better sustainability and higher income generation with special reference to locations. The government has made policies to encourage specialized breeders amongst the progressive farmers for commercial breeding. Species wise programmes were formulated and implemented. (Policy Note, SDAH, Rajasthan, 2011).

References

19th Livestock Census.(2012). Department of Animal Husbandry Dairying, and Fisheries.Ministry of Agriculture.GOI.

Andy L Offer. (1997). The Rural Business Network in the UK. First European Conference for Information Technology in Agriculture, Copenhagen, 15–18 June, 1997.

Annual Report.(2010-15). Department of Animal Husbandry, Dairying and Fisheries.Ministry of Agriculture.GOI.

ATIC interface meeting discussion paper. (2010). Directorate of Agricultural Extension.Indian Council of Agricultural Research.

Birthal, P.S. and Rao, P.P. (2002) Economic contributions of the livestock subsector in India. Technology option for sustainable livestock production in India.Procc. Of the workshop on Documentation, adoption and impact of livestock technologies in India, 18-19 Jan, 2001, ICRISAT- Patancheru, India:12-19

Chand, Sukumar.,AneeshSaraiya and V. Sridhar. (2010). Public Private Partnership in Indian Dairy Industry.Initiating White Revoution II.Technopak publication.

FAO. (2005). Case study- community based systems, Warananagar, India, Warana wired village project. Bridging the rural digital divide.FAO publication.

Gaikwad, Bhaskar. (2011). Three tier model for technology assessment, refinement and dissemination. Compendium of International Conference on Innovative Approaches for Agricultural Knowledge Management: Global Extension Experiences. Nov. 9-12; 2011

Ghotge, N.S., S. R. Ramdas, S. Ashalata, N. P. Mathur, V.G. Broome and M.L.S. Rao. (2002). A Social Approach to the Validation of Traditional Veterinary Remedies - The Anthra Project.Traditional Animal Health and Production,34: 121-143.

GOI (2009). Basic Animal Husbandry Statistic 2008. Department of Animal Husbandry, Dairying and Fisheries, Ministry of Agriculture, Government of India, N.Delhi.

Hegde, N.G. (2001) WHO challenges for Indian Dairy farmers. Yojana, December,:34-35,43.

Ian Christoplos. (1996). Poverty, Pluralism and Extension Practice. Gatekeeper Series No. 64.Sustainable Rural and Livelihood Programs.IIED.

Kathiravan, G., M. Thirunavukkarasu, R. Arumugam and C. Manivannan.(2009). Demand for public vs. private livestock services in South India: a double hurdle analysis. Indian Journal of Science and Technology. 2(2): 55-62

Mahapatra, R. 2012. Rise of livestock. Down to Earth. Tuesday 31st January 2012. Extracted from www.downtoearth.org.in on 6th October 2015

Nath, M., Elangoven, A.V., Mandal, A. B. and Johari T.S. (2002). Manual on 'MakeFeed' software for feed formulation. Published by Head, Division of Avian Nutrition and Feed Technology, Central Avian Research Institute, Izatnagar, 1-2

National accounts statistics. 2011. www.mospi.nic.in

Ozor, N. Agwu, A. E., Chukwuone, N. A. Madukwe, M. C. and Garforth, C. J. (2007). Cost-sharing of Agricultural Technology Transfer in Nigeria: Perceptions of Farmers and Extension Professionals. The Journal of Agricultural Education and Extension.13 (1) pp. 23-37(15)

Phand S, Tiwari, Rupasi and Sharma, M.C. (2009) Development of need based Animal Health Information System (AHIS) for veterinary professionals and students, Indian Veterinary Research Institute, Izatnagar. (Submitted for copyright).

Policy note. (2011). State Livestock Development Policy. Department of Animal Husbandry, Government of Rajasthan.

Policy Paper 39.(2007). National Council for Agricultural Policy Research (NCAP), ICAR, New Delhi.

Ramkumar, S., Garforth, C. and Rao, S.V.N. (2003). Information kiosk for dissemination of cattle health knowledge: evaluation report based on preliminary findings of research. (http://www.dfid-ahp.org)

Ravikumar S, Reddy K V R and Rao B (2007) Farmers' choice for cost recovery of veterinary services in different livestock holding systems- A case study of India. Livestock Research for rural Development,19 (5)

Sharma, M. C. and Rupasi Tiwari. (2011). Ensuring food security through livestock sector. Agriculture today year book. pp 102-108

Sree E. Karuna., S. Amarender Reddy and N. Veerabadra Rao. 2011. Transfer of technology clubs – a new approach for evergreen revolution in west Godavari district, Andhra Pradesh. Compendium of International Conference on

Innovative Approaches for Agricultural Knowledge Management: Global Extension Experiences. Nov. 9-12; 2011

Srivastava and Mehta (2014) Kisan SMS set to evolve. Accessed from www.lvemint.com on 9th October, 2015

State livestock development policy, Department of Animal Husbandry, Jaipur, Govt. of Rajasthan, http://india.gov.in

Thammi Raju, D. and Sudhakar Rao, B. (2006). An information technology enabled poultry expert system: Perceptions of veterinarians and veterinary students. *International Journal of Education and Development using Information and Communication Technology*.2(2) pp: 100-107. (http://www.iaita.ac.in)

Tiwari, R, Sharma, M.C. and Pachiyappan, K. (2012) Challenge specific extension models. Souvenier: Indian Dairying: Perspective 2020. XL Dairy Industry Conference held during 2nd to 5th February, 2012 at New Delhi: 65-69

Tiwari, Rupasi, Sharma, M.C. and Singh, B.P. (2010). Animal health information needs of livestock owners: A Kisan Call Centre analysis. Indian Journal of Animal Sciences. 80 (2):187-188

Tiwari, Rupasi. (2013) Modern Extesnion Teaching methods.In Bhushan *et al.*, eds. Livestock management (Production and economics). LPM, IVRI, Izatnagar pp: 351-368

Tiwari, Rupasi; Arya, H.P.S.; Singh, B.P. and Rawal, C.V.S (2005) "Remodelling extension strategies for livestock development" Indian Dairyman, Vol: 57 (3), Pp: 37 – 43.

Vijay Babu, D., M. Asokhan and M. Shanta Sheela.(2011). Perception of Dairy farmers on utility of Dairy information available through ICT. Compendium of National Seminar on Innovations in farming system research and extension for inclusive development, held at TANUVAS, Chennai. 24-25 Nov., 2011.

Vijayragavan, K. (2011). Innovative Extension Approaches and Models of IARI. Lead paper presented at National Seminar on Innovations in farming system research and extension for inclusive development, held at TANUVAS, Chennai. 24-25 Nov., 2011.

Wettasinha, Chesha. (2001). Scaling up participatory development in agricultural settlements. LEISA India, Vol. 3(3): pp.41-44.

Wijekoon, R.R.A. (2003). Concept paper presented to Sri Lanka council for Agricultural Research Policy.

Satish S and Premkumar N 1993 Are NGOs more cost effective than government in livestock delivery?A study of artificial insemination in India, London, Routledge.

Chapter 15

A Critical Review of the Extension Science as Practiced in India with an Agenda for Reform

Lalit Kumar Tyagi

ICAR-National Bureau of Fish Genetic Resources,
Canal Ring Road, P.O. Dilkusha, Lucknow – 226 002, U.P.

Though extension as a service, a profession, and also as a discipline, originated in the western world, India has had a rich heritage of early extension efforts in different forms of extension which later paved the way for evolution and development of nation-wide extension programmes. Over the years a wide national extension system has evolved in the country which has gone through a series of reforms with a purpose to transform it into a vehicle for growth-oriented agricultural and rural development of the country. Extension as a discipline has also grown over the decades and today it is considered as an applied social science!. A large no. of M. Sc and Ph. D students come out of various universities in Extension with different discipline names. A plethora of research studies, cutting across a wide range of target groups, topics and geographical areas of the country are carried out and a large volume of dissertations/reports generated through these studies. Number of seminars/workshops/symposiums and training programmes are also organised by the extension scientists. All these efforts have produced a large body of facts/knowledge, resulted from observation and investigation, generalized into principles and theories, which could be useful for extension educators.

Despite these efforts, some uneasiness extension scientists (particularly among young scholars) is felt towards their discipline both at intellectual, as well as, professional level. However, not everyone shares this uneasiness! There is indifference too! This is because Extension has always been in hard times to be recognized as a science and a discipline at par with other biological and social sciences in India! Extension is still considered anybody's job! and it is not uncommon to find extension scientists as poor members of technology development teams! The purpose of this paper is to transfer this uneasiness and indifference into issues to be discussed and thought through by bring it on a single platform. The paper discusses in detail: 1. administrative and image-related issues, 2. issues related to the teaching and curricula, 3. issues related to research and 4. issues related to professionalism in Extension as a science in India. Drawing upon recent literature and author's interactions with a wide variety of extension professionals across different forum and sectors, the paper also comes up with an agenda for reforming/reorienting Extension Science in India.

Introduction

The 'extension' first originated as "University Extension" in Britain in the 1840s. Mr. James S. Stuart approached Cambridge University to organise centers for extension lectures. Cambridge University formally adopted it in 1873, followed by London University and Oxford University. By the 1880s it was referred to as the 'extension movement'. 'Agricultural extension', however, originated in the United States of America in early 20th century and was strengthened in the form of Cooperative Extension Service. Since then it has evolved and referred to various names and defined in various ways in different countries using various elements in different combinations (Pl. see Singh 1973, Dhama and Bhatnagar 1985 and van dan Ban 1995 for details). Broad common meaning of 'extension' has been communication of relevant information and technology to help people rural people form sound opinions, make good decisions and improve their livelihoods, mainly in agriculture and allied sectors. Currently, 'the term 'extension' has become a generic term employed to refer to the variety of systems and providers that have emerged for communicating and transmitting information and technology to farmers and other rural populations. Extension is often viewed as comprising public, private and semi-public 'systems' that make up a multi-sectoral 'pluralistic' system' (Rivera and Sulaiman, 2009).

Evolution of Extension in India: From Work/Practice to Teaching

India has had a rich heritage of early extension efforts in different forms including various sporadic attempts in pre-British period, efforts by the British Government, efforts by Indian reformers, voluntary organisations and state governments and rural reconstruction work by Christian missions. Experiences gained in these efforts led to the need for intensive extension efforts at national level which resulted in the form of the Community Development Programme launched in 1952 culminated into the National Extension service in 1953. This was the turning point in the history of extension in India. This meant that a need was felt for systematic education and training of personnel for agricultural and community development. As a result, agriculture teachers trained by U.S. extension professors started teaching extension courses in agriculture colleges. Thus, 'extension' in India started from work/practice and later evolved into a subject for teaching and training. Over the years a wide national extension system has evolved in the country which has gone through a series of reforms with a purpose to transform it into a vehicle for growth-oriented agricultural and rural development of the country.

Growth of Extension Discipline in India

Extension as a discipline has had its own share of problems right from the beginning. from its humble beginning as work/practice, evolved into a discipline in India. Initially, there was lack of teachers trained in extension to teach and train. Therefore, courses were taught by people trained in other disciplines of agricultural sciences. Dhama and Bhatnagar 1985 noted, "In the beginning, the mushroom growth of extension education in the form of institutions and organizations, a highly heterogeneous group of professionals and the teaching of the discipline by unskilled and untrained staff hampered the professional growth of the discipline in India. But, the untiring efforts, zeal and enthusiasm of the academicians of the discipline who demonstrated their studies and researches and virtually 'educated the educated masses', were able to bring the discipline on par with other well recognized ones". Growth of extension as a discipline in India started with starting of post-graduate teaching in the mid-fifties at Bihar Agriculture College, Sabour, followed by Ph.D in Agril. Extension started at Indian Agriculture Research Institute (IARI), New Delhi in 1960. In 1960s several developments took place which helped in the extension taking off as a discipline in India. This was the period when a chain of State Agricultural Universities (SAUs) started with assistance from U.S.A. Extension was taught as a subject for UG students in agriculture, veterinary, etc. During this period, PG programme in extension (later on Ph.D also) was started in SAUs and several agricultural colleges. The momentum was boosted with the formation of Indian Society of Extension Education, and launch of World's second journal of extension, Indian Journal of Extension Education at IARI, New Delhi in 1964 and 1966, respectively. Later on (1970s onwards), Departments of Extension and Continuing Education were started in general universities with varying names. This growth was further strengthened when some state chapters/societies of extension education started (*e.g.* T.N., Maharastra, Rajasthan, etc.,), and later on few other professional bodies like International Society of Extension, Society of Extension Education, Society for Community Mobilization and Sustainable Development also came up. Many of these societies also started publishing their separate journal.

Extension, thus, as a discipline has also grown over the decades and today it is considered as an applied social science! A large no. of M.Sc and Ph.D students come out of various universities in extension with different discipline names. A number of seminars/workshops/symposiums and training programmes are also organised by extension scientists. All these efforts have produced a large body of facts/knowledge, resulted from observation and investigation, generalised into principles and theories, which could be useful for extension educators.

Evolution of Extension Research in India

A plethora of research studies, cutting across a wide range of target groups, topics and geographical areas of the country are carried out in extension and a large volume of dissertations/reports are generated through these studies. Extension research in India has also gone through a process of progressive evolution over last six decades.

It started as 'exploratory research' in late 1950s with focus on 'what is where'; moved on to 'descriptive research' where focus was on 'what is what' which included observational methods, case-study methods and survey methods and further expended to 'explanatory research' in 1970s which stressed upon 'cause and effect relationship' and statistical analysis such as chi-square and correlation became popular among the researchers during this period. Increased use of 'experimental research' was witnessed in 1980s with focus on knowing how much effect was caused by how much change in causal variables, by manipulation of an independent variable and control of extraneous variables. This decade of extension research witnessed more sophistication with regards to statistical analysis. Regression, path analysis, factor analysis were used to develop a better comprehension of the way extension variables were operating in a field situation. A number of studies focussed on understanding the social system rather than variables in isolation. The relationships of extension variables in a system and their operating behaviour could be estimated by now. The 1990s, however, witnessed a paradigm shift from quantitative research to qualitative research. Importance of Indigenous Technical Knowledge (ITK), participatory research, etc. was felt. Adoption studies, however, continued to be the major area of research. With the advent of twenty first century, a number of new research areas, such as cyber extension or e-extension, privatisation of extension, entrepreneurship development, group farming, Information and Communication Technologies (ICT), etc., found place among extension researches (Adapted from Kumar, 2015).

Today, a vast professional force of extension researchers exists in the National Agriculture Research System, as well as in other organisations like general universities and Non-Government Organisations (NGOs). Extension researchers are placed in various sectors at different parts of the country with varying socio-economic and agro-ecological landscapes.

Why this Paper? - The Focus

Despite these efforts, some uneasiness extension scientists (particularly among young scholars) is felt towards their discipline both at intellectual, as well as, professional level. However, not everyone shares this uneasiness! There is indifference too! This is because, 'in the universities, extension education has always had hard time to be recognized as a science and a discipline at par with other biological and social sciences' (Sulaiman and van den Ban 2000).

The purpose of this paper is to transfer this uneasiness and indifference into issues to be discussed and thought through by bringing them on a single platform. The paper discusses: 1. image-related issues, 2. issues related to the teaching and curricula, 3. issues related to research and 4. issues related to professionalism, in extension as a science in India. Drawing upon recent literature and author's interactions with a wide variety of extension professionals across different forum and sectors, the paper also comes up with an agenda for reforming/reorienting 'Extension Science' in India.

Issues and Challenges in Extension Science in India

Image-Related Issues

Extension as a scientific discipline as well as a profession has suffered a lot due to its poor image among other disciplines in the National Agricultural Research System (NARS). Traditionally, many senior extension management positions in NARS have gone to members of other disciplines and it has been an issue of concern among the extension professionals. There has been lot of improvement in this regards, off late, though. However, the situation down the line amongst colleagues is still 'challenging' for young extension scientists. The typical attitude towards 'extension' continues to be like 'yeah, its a useful work/service for taking technologies to the farmers but what research extension scientists would do'? The typical images in which extension scientists are perceived mainly include: 'men of transfer of technology (TOT) for TOT!, media men! good event organizers/manager!, exhibitors! or at the most, surveyors!, etc,. The typical mindset that 'extension is anybody's job!' still continues, albeit with reduced vigour then before. Writing about extension scientists in various research institutes in India Sulaiman and van den Ban, 2000 note, "their status has been generally inferior, their potential role as independent researchers or as a member of technology development teams is not given adequate importance. Seldom have they played a role in developing a strategy for two-way communication between researchers and farmers". Many times, the extension scientists could be found as poor members of technology development teams!

While searching for reasons for this state of affairs, however, blaming the system or other colleagues for this poor image wouldn't serve the purpose. Rather, introspection seems a better way to look at it. Sulaiman and van den Ban, 2000 had put it in this way, "historically these conflicting role perceptions for social scientists existed in the agricultural research institutions all over the world. But over the years the social scientists in the international agricultural institutes, comprising mainly economists, sociologists and anthropologists, have slowly established their identity and usefulness and convincingly earned the respect which they deserve. However, this has not happened in India. Those at the helm of affairs are still skeptical of the role, the social scientists comprising mainly of economists and extension scientists) can play in the National Agricultural Research System (NARS). The poor institutional support, divergent role expectations and the week training in research methodology have adversely affected the morale of the extension scientists. Lack of a clear understanding of the role of extension professionals in the NARS have also contributed to this situation. Extension professionals also seem never to have articulated their potential contributions very effectively".

Extension scientists thus are, in general, a demoralized lot amongst their colleagues in NARS! This statement sounds too simplistic and generalized an expression, and off course, can not apply to everyone out there. There could be, and there are, several extension scientists doing very good research and making their usefulness and significance felt in the system. However, this is used here symbolically to indicate the prevailing general environment in the professional circles. Sivakumar, 2015 has articulated this issue meticulously. He writes 'extension scientists are often under pressure to deliver tangible outputs to show impact of technologies whereas the poor technology adoption is often the results of a number of factors including the weakness in field extension'. 'Many of the research managers have a different perception on the role of extension scientists. Many of them believe that the role of extension scientists is to promote technologies (developed at the respective centres) in the field, rather than contributing to the effectiveness of the research projects through research and publications'. These perceptual differences have created a stressful working environment where the extension academicians/scientists are overburdened to meet the immediate goals of their supervisors leaving aside their research ambitions' (Sivakumar, 2015).

There could be several other reasons too. There seems to have been a 'confined to our shelves' attitude among a majority of the extension professionals! Lack of initiative by the senior extension professionals to break the myth and to change the image of extension by strong lobbying at the top, lack of seminal research works with strong theoretical and/or practical footing so as to yield compelling findings with sound policy and planning implications, adherence to the conventional mindset within the senior extension fraternity about: research topics and approaches and linkages with other social and biological sciences, and failure to promote enthusiastic young scientists to infuse innovations and enrich the subject to establish its identity like other social scientists in international agricultural institutes, etc, are some of the reasons which could have also contributed to the poor image of the discipline! Though, some change in perceptions seems to have taken place in recent times due to good work carried out by extension researchers at various places, still a lot remains to be done in this regard.

Issues Related to the Teaching and Curricula

As was mentioned above, teaching in extension started as a felt need to train people for community development/national extension service/agricultural development work. This history has had its impact on the way course curricula have evolved over the years. Also views on extension have also been changing over time (pl. see Sulaiman and Hall, 2002; Rivera and Sulaiman, 2009 and Sulaiman, 2011 for a detailed discussion on it) which have thrown new challenges to the discipline. Sulaiman and van den Ban, 2000 had discussed these issues in detail with suggested strategy for reorienting curricula. Though their paper was written over one and half decade ago, most of the issues raised by them are still relevant even today! We are still producing 'three-in-one generalists' at Master's level (who could be equally 'competent' or 'in-competent' for being an extension researcher or extension service worker or a development media professional!). Lack of match between course curricula and changing job market requirements; lack of job opportunities for extension students; deficiencies in curricula and teaching approach; mere learning of basic concepts and definitions of related other social sciences, *e.g.* sociology, psychology, management, etc.; inadequate opportunities to understand and analyze complex rural social realities and development processes and lack of integration of the developments in other social sciences in the extension curricula, jumping on to new concepts/jargons emerging in the international extension and development field without thinking through or required efforts to relate to them or finding out ways to put them into practice in actual practical situations in the country, etc, are some of the issues which continue to hamper the growth of the discipline in the country. Reorientation/redesigning of curricula have been discussed, attempted but not much changed except few minor innovations. Of late, there have been attempts to introduce few innovative courses and topics at various places within the given course structure, however, lot remains to be desired in this regard.

Issues Related to Research

Research is needed in any discipline to create knowledge which fuels the growth of the discipline. As mentioned above, though extension research has travelled a long journey, it has been criticized for suffering from several lacunas. 'People always expect extension professionals to "do" something which will help the farmers in a direct way, rather than indulging in "genius" things like conducting high quality research and publishing papers, etc. When technologies are not adopted or farmers are not willing to take up research outputs, it is regarded as "extension failure". For a scientist/academician from other disciplines, scientific publishing in extension is a kind of useless activity' (Sivakumar, 2014). Some of the issues and challenges extension research is facing are listed below:

1. Mainly student's researches: lack of focused research projects from extension faculty for substantial output. 'Requisite resources for research in extension are always in short supply due to lack of clarity between extension research and extension work. Lack of capacity to formulate and submit multi-disciplinary research projects under competitive grants also adds to this challenge' (Rao *et al.*, 2014).
2. Extension research also suffers from imbalance in its coverage:
    a. Topics and issues covered
    b. Geographical areas and target groups covered
3. Extension research mainly focused on preoccupation with *'troubles'* - which occur within the individual and within the range of his immediate relations with others; they have to do with his self and with those limited areas of social life of which he is directly and personally aware. It has given inadequate little or inadequate attention to *'issues'* - which deal with matters that transcend these local environments of individual and the range of his inner life and have to do with the organization of many such milieu into institutions of an historical society as a whole, with the ways in which various milieu overlap and interpenetrate to form the larger structure of social and historical life. An issue often involves a crisis in institutional arrangements and cannot very well be defined in terms of the immediate and everyday environments of people. Two-way interaction between studies of *'troubles'* or *'milieu'* of the individuals and studies of *'issues'* or *'structure'*

has been missing in most studies. We can see this in most of the extension studies focusing on the 'subject' of the extension *i.e.* farmers, while neglecting other aspects of the system.

4. Extension research seems to be caught in 'paradigmatic syndrome': A paradigm denotes some accepted examples of scientific practice – examples which provide models from which spring particular traditions of scientific research. Through some achievements and promise of providing 'successful solutions' to future problems, one school/paradigm 'emerges triumphant' and is able to attract a large and enduring group of adherents. Extension research has suffered due to its preoccupation with the TOT paradigm! Though the extension science globally has moved beyond technology transfer to facilitation, learning, organising and building networks, extension research in India is still stuck in studying technology transfer (Prasad, 2013). Though the research in extension has wider scope, it mainly focuses on adoption and diffusion of innovations and often revolves around research – extension – farmer linkages. Extension research has many such self-imposed limitations (Rao *et al.*, 2014).
5. Obsession of extension research with the TOT paradigm has resulted in lack of interdisciplinary inquiry taking changing views and emerging new/ broader roles of extension into account. Its consequence has been lack of impact making researches on felt needs having strong policy significance.
6. Poor professional standards in conducting research are another area of serious concern in extension. Prasad, 2013 has meticulously summarized them.
7. There have been several methodological issues in extension research: We find over-prevalence of routine studies of which many rarely go beyond demographic, socio-economic and perception – related variables, with little attention to identification of clear research questions, inadequate definition of problems, ad-hoc selection of variables, over/or under-emphasis on quantification, mechanical description and interpretation of results, etc.
8. Methodological advancement lacking: 'While research tools and techniques in core disciplines from which extension borrowed its research methods have evolved significantly, extension research still depends heavily on many of the outdated tools' (Sivakumar, 2013). 'Current research tools in extension are outdated and their continued use has resulted in stereotypic and insignificant outcomes' (Sivakumar, 2015).
9. Further, due to inadequate training in research methodology, there is lack of clarity and competence to apply various research approaches (quantitative, qualitative, action research, etc,) appropriately depending upon the research question to be addressed. Lack of competence in statistical methods, tools and techniques among the extension scientists, only makes their task tougher.
10. Lack of theory construction has been a major hindrance in growth of extension discipline in the country. Due to lack of practical significance

and stagnation in theory development, extension research is often criticized by other agricultural disciplines as a "non-performing discipline" (Sivakumar, 2015). 'While several other disciplines consider theories as the pillars on which their subject matter is built upon, there is a widespread opinion among extension scientists and professionals that extension should be "applied" in nature and therefore its focus should not be on developing theories and models that are purely academic' (Sivakumar, 2013). There are two issues involved here: First, because of its 'applied social science' tag, basic research has been 'considered' as a 'taboo' in extension research! Second, there has been virtual absence of efforts to compare, contrast, synthesize and integrate studies to the extent possible for theory construction and conceptual advancement of the discipline based on empirical evidences. This issue is also considered and discussed in the form of 'meta analysis in extension' and is referred to as 'conducting research about research' (Prasad, 2014). Meta analysis 'refers to the methods that focus on contrasting and combining results from different studies, in the hope of identifying patterns among study results and sources of disagreement among those results. A meta analysis gives thorough summary of several studies that have been done on the same topic/theme and provides the reader with extensive information on whether an effect exists and what 'size' that effect has. The main advantages of meta analysis are that the results can be generalized to a larger population and the precision and accuracy of estimates can be improved as more data is used' (Prasad, 2014). Such studies have been done by various scientists in different social science disciplines/field of studies globally, *e.g.* environmental economics, fisheries management, and many more. There are, however, practical constraints too in venturing into this methodologically highly demanding field as 'it requires deeper understanding of advanced statistical tools' (Sivakumar, 2014).

Professional Issues in Extension

1. Inadequate and poor quality of publications in extension: Quality of publications is very poor in extension. Mainly routine data based ('table based'!) papers are given weightage where as issue-based discussion/ review/concept papers are negligible. Similarly, original, innovative books are not forthcoming. Poor quality of publications in extension is a manifestation of several interrelated factors. These include poor quality of research output generated due to lack of clarity on the role of extension scientists, inadequate training in research methodology, lack of publications from several practitioners who are engaged in field extension activities, lack of capacity to prepare quality research papers due to limitations of drafting skills, articulation of issues, language constraints, poor quality of review and editorial decisions of the submitted papers, etc. Rao *et al.*, 2014 has discussed this issue in detail.

2. Another problem is 'limited presence of extension research in reputed peer-reviewed international multi-disciplinary journals such as Agricultural Systems, Outlook on Agriculture or Rural Sociology as well as international extension journals such as Journal of Agricultural Education and Extension and Journal of International Agricultural and Extension Education. Even those who are engaged in quality research in extension pay very little attention to publishing their research outputs in some of these reputed journals' (Prasad, 2013).
3. Few extension scientists in ICAR Institutes and KVKs are doing extremely good research work. However, their expertise is relatively under-utilised by the SAU research system. Unlike IITs, IIMs and many engineering and science colleges, the SAU extension research did not allow their students to get an outside research advice or collaboration (Sivakumar, 2014).
4. Formal constructive criticism and peer discussion has virtually been absent in the country. Few initiatives have, however, emerged off late, which are mentioned in the next section.
5. Informal quality professional personal communication is in its infancy.
6. Professional attitude is sadly missing among the extension scientists.
7. Efforts to make the work of extension scientists available to real users and to sell the discipline and its products are lacking.
8. Despite growth in numbers in last two decades, inadequate interaction and lack of professionalism among national/state/regional professional societies of the discipline has also not been able to help the cause of professional growth of the discipline as could be expected.
9. Seminars/symposiums have become mere rituals! They are often not issue based with definite purpose and outcome, and results in routine papers presented often without any relevance to the theme of the seminar (For a very good discussion on this topic, pl. see Prasad, 2013 and Kumar, 2014).

Where do we go from here: Agenda for Reforming/Reorienting Extension Science in India?

Above review gives us enough pointers as to what needs to be done to resolve the issues and overcome the challenges extension as a scientific discipline is facing in the county. Concerned professionals have been articulating several of them. Based on the review and authors' own views, major aspects of the agenda for reforming extension science are summarized below:

- Present a strong case in NARS for preventing senior extension management positions from going to the non-extensionists and actively and aggressively pursue it.
- Resolve the confusion/conflict over names of degrees in Extension.
- Being 'in touch' with recent happenings to be alert enough so that assignments which require extension expertise don't go to others.

- ✫ Need to come out of our self-imposed shelves, open up ourselves and develop more cordial and fruitful linkages with other social and biological sciences. This, however, would require concerted efforts both at the personal level of scientists and at professional fraternity level, towards updating ourselves and capability development.
- ✫ Individual as well collective efforts (particularly by senior extension scientists/professors) to break the typical image/attitude/mindset about extension prevailing in NARS.
  - Educating and advocating our counterparts about our potential contributions to improve their woks,
  - Taking active part in interdisciplinary seminars, meetings, etc. and identifying gaps/avenues where contributions from our discipline are required to make others' work useful, showing the difference we can make,
  - 'Assertiveness' as a member in technology development teams.
- ✫ Need to win over the conventional mindset within the discipline to follow the established path of conducting 'routine extension researches' and encourage and promote young scientists to dare to innovate and raise the discipline.
- ✫ Reorienting the course curricula at each level: As articulated by Sulaiman and van den Ban, 2000 and others and also discussed in various seminars/meetings, serious restructuring, need based specializations, focus on skills and competencies development and assimilation of concepts and updating with recent advancements from other social sciences but with a clear understanding of how to apply and adapt (*i.e.* 'put them to use') to suit real extension/rural development situations, etc, need immediate attention and action.
- ✫ Customizing research, *i.e.* focusing on few research themes and develop "methodological" expertise to achieve "tangible" and "visible" outputs is the need of the hour. To do this, we need a significant change in the extension streams at PG level. At the M.Sc. level, we can offer extension courses with specializations in ICTs, Livelihood system, Media communication, Entrepreneurship, etc. If we define our universe with certainty, the research under each stream will be specialized and yield some tangible output. The graduating students will also feel like they acquired something new and valuable (Sivakumar, 2014).
- ✫ There is need to substantiate the discipline by production of quality research work of wider and deeper policy and practical relevance. Therefore, extension research needs an all round reorientation and improvement: in broadening of its coverage, improving its rigour by methodological advancement and capability development, and raising professional standards.
- ✫ A lot needs to be done to improve training on research methodology at the post-graduate and doctoral levels. New research methods (both quantitative

and qualitative) are yet to find a place in extension curricula. The students and faculty also do not have access to latest books and journals due to budgetary restrictions in academic institutions. Low budgetary allocation to social science text books and journals is also a reflection of the poor image of extension discipline among administrators and others who take these decisions (Rao *et al.,* 2014). Recently, Sivakumar in his blogs (2013 and 2015) has elaborated nicely on availability of new tools and their applicability for extension researchers, and suggested a practical and pragmatic approach for extension scientists to conduct significant and useful research amidst existing constraints.

- ✰ Extension research needs to move beyond TOT and diffusion – adoption paradigms. Diversification of research areas is needed to respond to the changing circumstances and needed of the stakeholders in changing socio-economic and politico-economic local, national and global scenario. Identification of critical research programmes and a series of focused research studies to come out with useful findings in a particular field of study to solve a critical/felt need of the stakeholders would greatly enhance the image and practical significance of the discipline among peers from other disciplines and also policy makers. There have been some attempts in this direction which have enhanced the relevance of extension research. However, they are few and far between. Greater efforts and manpower need to shift attention towards this direction from the current practice of conducting routine extension studies focusing around demographic profile, 'felt needs', etc.

  There are several institutional innovations, roles and themes in extension systems/work which are emerging from the field out of pressing needs. Similarly, there are various advancements taking place in other relevant fields of studies. Extension research needs to broaden its outlook and cater to the needs of these changing demands as well as new opportunities for contributing in improving the rural livelihoods (for a good discussion on it pl. see Prasad, 2014 and Gowda et.al 2014). There is also thought gaining strength that 'extension should play a pivotal role in influencing policies too rather than focusing only on technology dissemination as the former has strong influence on technology dissemination' (Rao *et al.,* 2013).

- ✰ Research methodology is another field where a lot needs to be done: Interaction, blending/integration of the studies of troubles/milieu and the studies of structure/issues is needed so as to get the proper perspective on the problem studied and complete the picture. Greater attention need to be given to clear identification and articulation of research questions, adequate definition of the problem, appropriate use of research method and tools, and interpretation of findings in terms of research questions.

- ✰ Regular efforts are needed to compare/contrast/review and synthesize research studies in identified research areas to come out with theoretical underpinnings, as well as, practical implications. This, however, would require sound understanding of advanced statistical tools, so as this becomes an objective exercise.

- The concept of 'research' in extension needs to be broadened, recognizing that beyond the public research and extension organisations, a range of actors have important and vital roles in the generation and dissemination of agricultural innovation (Prasad, 2014). Many extension practitioners are involved in extension work but they rarely publish research articles, may be due to their lack of need/interest for such publications or capacity to write articles from their work experience. It is a big loss to the extension profession. It is important to harness such useful experiences in different crop and livestock systems and make them accessible to others through publication (Rao *et al.*, 2014).
- There is urgent need to improve professional standards in various ways. Some of these ways have been nicely articulated by several professionals in recent blogs (Rao *et al.*, 2014; Prasad, 2013, Kumar, 2014 and Sivakumar, 2015). Important points requiring serious attention are summarized below:
- Greater interaction and better professional standards among different professional societies is needed. Professional societies need to mature and transform them 'to play new roles relevant for the future of the discipline. These include enhancing skills related to: Formulating good projects, employing various research tools, Identifying new areas of research, Writing research articles – presentation skills, and Accessing new research grants' (Rao *et al.*, 2014).
- There is need for setting a tradition of, and mechanisms for, formal constructive criticism and peer discussion. Off late, a renewed interest is clearly visible in this regards through increasing professional discussion among peers through new emerging professional networks, blogs and social media. Portals like Agricultural Extension in South Asia (AESA), initiatives like recently launched 'Indian Extension Network' and increasing sharing and discussion by extension professionals on matters of professional growth via social media like facebook, are examples of this growing welcome trend. The need is to get involved in such initiatives and contribute by participating in discussions and knowledge sharing for professional advancement of the discipline.
- Individual efforts to enhance personal professional communication to enrich each other and inculcate professional attitude are also needed.
- Systematic efforts to increase our visibility and sell our work, skills and products are need of the hour. This, however, again, would call for 'honing' our skills and competencies by updating ourselves.
- Much needs to be improved in organizing seminars/symposiums – with definite purpose and outcome with strict scrutiny of papers. However, focus on quality, not quantity should be a precondition, if they are to make any impact on the professional growth of the discipline.

- ✰ Small/compact issue-based brainstorming sessions/workshops by invited relevant experts, aimed at specific outcome, with selected background/lead papers, no formal presentation of number of ritualistic, irrelevant, old 'facts' cluttered 'papers', a clear focus on detailed and intensive discussion on points raised in the background/lead papers, with a definitive outcome in terms of policy guidelines or action points for practitioners, could prove highly impact-making in enhancing the image of the profession.
- ✰ Quality of publications needs to be improved urgently. This is a huge task in itself. Something which calls for action on several fronts including shedding our inhibitions, resisting our temptations for short-cut in publishing poor papers, giving preference to quality over affiliations, strict scrutiny of papers, training/guidance in good paper writing, due importance to be given to concept/review/discussion papers, improving the quality, sincerity and incentives for peer reviewing, etc. If we take care of these issues, rating of extension journals will also improve.
- ✰ Also specific research/review based concise publications on topical issues of policy and practical relevance and original/innovative books to enrich the subject and make it more interesting, are a pressing need of the discipline.

Conclusion

In the background of the evolution and growth of the extension science as a discipline, this paper has discussed some of the issues and challenges it is facing in the current professional environment in the NARS. An agenda for reforming the discipline by much needed actions on various fronts has also been outlined in the paper. It is hoped if we could improve upon the indentified points, extension could retain its due place as a vibrant scientific discipline in the country.

References

Babua, Suresh C.,P. K. Joshib, Claire J. Glendenningc, Kwadwo Asenso-Okyerec and Rasheed Sulaiman V. 2013. The State of Agricultural Extension Reforms in India: Strategic Priorities and Policy Options. *Agricultural Economics Research Review.* 26(2):159-172.

Gowda C M.J., Dixit S, Burman R and Ananth P N. 2014. Extension Research and Technology Development Blog 29, *Agricultural Extension in South Asia.* (available at *http://aesa-gfras.net/ Resources/file/FINAL-M_J_Chandre per cent 20Gowda-13-FEB.pdf*).

Kumar, Suresh 2014. Is it time to organize a seminar on how to organize seminars? Blog 30. *Agricultural Extension in South Asia.* (available at *http://aesagfras.net.in*).

Kumar, G.A.K. 2015. An overview of Extension Research in India. *Personal Communication.*

Prasad 2013. Research in Extension: It is time to introspect. Blog 5, *Agricultural Extension in South Asia.* (available at http://aesagfras.net/Resources/file/Blog per cent 205Research per cent 20in per cent 20 Extension per cent 20It per cent 20is per cent 20time per cent 20to per cent 20introspect.pdf).

Prasad 2013. My Meeting Notes. No.7/18, *Agricultural Extension in South Asia.* (available at *http://aesagfras.net.in*).

Prasad 2014. Extension Research: Random thoughts from a Well Wisher. Blog 39, *Agricultural Extension in South Asia.* (available at http://aesagfras.net/blog.aspx?id=98 and title=Extension per cent 20Research: per cent 20 per cent 20Random per cent 20thoughts per cent 20from per cent 20a per cent 20 Well per cent 20Wisher).

Rao, S.V.N., K. Natchimuthu and S. Ramkumar 2013. Beyond technology dissemination: Why extension should also focus on policy change?. Blog24. *Agricultural Extension in South Asia.* (available at *http://aesa-grafs.net/Resources/file/Blog per cent*).

Rao, S.V.N., K. Natchimuthu and S. Ramkumar 2014. Scientific publishing in Extension: Are we doing enough and are we doing well?. Blog 40. *Agricultural Extension in South Asia.* (available at *http://aesa-grafs.net/Resources/file/Blog per cent 2040.pdf*).

Sethurman Sivakumar 2013. Research in Extension: New tools to reinvent its future. Blog 4, *Agricultural Extension in South Asia.* (available at *http://aesagfras.net/Resources/file/ Blog per cent 204 per cent 20Enhancing per cent 20the per cent 20potential per cent 20of per cent 20quality per cent 20videos per cent 20for per cent 20farmers.pdf*).

Sethurman Sivakumar 2014. Comments on Prasad 2014. Extension Research: Random thoughts from a Well Wisher. Blog 39, *Agricultural Extension in South Asia.* (available at http://aesagfras.net/blog.aspx?id=98 and title=Extension per cent 20Research: per cent 20 per cent 20 Random per cent 20thoughts per cent 20from per cent 20a per cent 20 Well per cent 20Wisher).

Sethurman Sivakumar 2015. Negotiating Reality: A pragmatic approach for conducting quality extension research. Blog 44, *Agricultural Extension in South Asia.* (available at *http://www.aesa-gfras.net/Resources/file/RS-eds-12FEB-2015 per cent 20final per cent 20ver per cent 20Blog per cent 2044 per cent 20(1).pdf*).

Sulaiman V Rasheed 2011. Extension-plus: New Roles for Extension and Advisory Services. *In:* Agricultural Innovation Systems: An Investment Sourcebook.

Sulaiman V Rasheed 2012. Agricultural Extension in India: Current Status and Way Forward. Background Paper prepared for the Roundtable Consultation on Agricultural Extension, Beijing, March 15-17.

Sulaiman V Rasheed and A.J. Hall 2002. Beyond technology dissemination: reinventing agricultural extension. *Outlook on Agriculture*. 31(4): 225-233.

Sulaiman V Rasheed and Anne W. van den Ban 2000. Reorienting agricultural extension curricula in India. The Journal of Agricultural Education and Extension. 7(2): 69-78.

William M. Rivera and Rasheed Sulaiman V. 2009. Extension: object of reform, engine for innovation. *Outlook on Agriculture*. 38(3): 267-273.

Chapter 16

An Assessment of Public Private Partnership Extension Approaches and Scope for ICT Platforms in ATMA

K.N. Krishnakumar[1], Anup Prakash Uppadhyay[2] and Kumari Manisha[3]

Faculty of Communication and Extension Management, Indian Institute of Forest Management, Bhopal, M.P.

A general weakness of agricultural extension system in developing countries is the low adoption of new agricultural technologies by farmers at lower end. The situation becomes even worse when it comes to adoption of Information and Communication Technologies. The importance of agricultural extension in transferring relevant knowledge and information to farmers as well as in translating policy directions into action is well known. As the present public extension systems in India have not yet become self sustained to effectively address the issues and concerning with small and marginal farmers, a search for new extension models that are more effective, efficient, and responsive to different categories of farmers is required to be explored and developed. The exploration of new approach has been started as a part of reform process in agricultural extension system initiated when Agriculture Technology Management Agency project was launched in seven states *viz.* Andhra Pradesh, Himachal Pradesh, Bihar, Jharkhand, Maharashtra, Orissa and Punjab as an autonomous body under the National Agriculture Technology Project in the year 1998. The reform in extension system was felt necessary since long time, because the conventional system was primarily top-down in nature and implemented with negligible participation of farming communities in the technology transfer process which was primarily dominated by the agenda set by the mostly govt. owned extension agencies. The feedback process in the traditional extension system was also very weak and real-time field problems were not addressed by researchers and planners because of lack of objective- information. The research and extension recommendation were too general and blanket like, covering entire agro climatic zone irrespective to needs of the farmers. Alternatively, with the active involvement of private stakeholders in agricultural sector to meet demands of the farming community, a Public-Private Partnership concept, focusing on the partnership between the private sector and government extension bodies emerged to promote extension efforts more efficiently. This partnership has emerged as one of the crucial strategy in agricultural extension in the form of ATMA. Currently a large number of ATMAs have developed partnerships with the private sector bodies in the processing industry, farmer's organizations, cooperatives, corporate bodies etc. This partnership has facilitated the dissemination of technologies, the supply of quality inputs (seed, micro-nutrients, bio-fertilizers, bio-pesticides, technological tools etc.) and marketing of farmers' produce. In addition, ATMA plays a crucial role in building the managerial capacity in the public sector as it is has to coordinate and regulate the functions of private sector while rendering extension services to the farmers, especially those who have small and marginal land holdings. This paper explores the grounds on which Public-Private- Partnership model is working and the challenges associated along with the possible suggestion. The paper also discusses the possibilities of ICT based platforms to provide high-value information support to the farmers with the support of PPP model. The study findings suggest that the modern and information rich extension system requires greater integration of different key stakeholders through inclusive-models like that of PPP at the multi levels of implementation. At the same time, the extension staff at those levels needs systematic training in areas like planning, resource management, monitoring, evaluation and documentation. Also, the incorporation of ICT requires capacity building and infrastructure support.

Keywords: *Agriculture extension, Agriculture Technology Management Agency (ATMA), Strategic research and extension plan, Public private partnership, ICT.*

Agriculture is the major occupation in India, supporting 73 per cent population directly and contributes substantially to the national economy as well. About 1/3 of the total gross domestic product comes from agriculture sector alone. In export front agriculture has been contributing more than 11 per cent of the total export. The Indian farmers have made tremendous progress in the last three decades and today they are self sufficient to feed one billion populations besides sizable exports. Indian agriculture is mainly monsoon dependent with limited land resources.

Agriculture provides a livelihood for more than 60 per cent of developing country population, and in many countries, farm families make up 80 per cent or more of the population. Agriculture has already reached the limits of land and water, thus future increases in food production must exploit biological yields on existing land **(World Bank, 1997)**. The World Development Report 2008 **(World Bank 2008)** emphasizes agricultural extension as an important development intervention (i) for increasing the growth potential of the agricultural sector in the light of rising demand-and supply-side pressures, and (ii) for promoting sustainable, inclusive, and pro-poor agricultural and hence economic development. The field of extension encompasses a wider range of communication and learning activities including agriculture, agricultural marketing, health, and business opportunities etc. The arrangements for agricultural extension in India have grown, over the last five decades, in terms of activities, organizational types and available manpower. The degree of institutional and economic development determines the scope and types of agricultural extension services and the ways in which these services are provided and financed (Anderson 2007). In a developing country like India extension education has become one of the arms of the national governments to promote agricultural development.

A major change in public sector extension came with the implementation of the World Bank sponsored Training and Visit System (T and V) in 1974. Most States adopted the T and V system during the 1980s, and this improved the financial and human resource capacity of the extension system. The 1970s also witnessed the launch of Krishi Vigyan Kendras (KVKs) or Farm Science Centres, Lab-to-Land programmes, and Operational Research Programmes by the ICAR. Krishi Vigyan Kendras (KVKs) were begun by ICAR to provide need-based and skill-oriented vocational training to farmers, field-level extension workers and other self-employed persons. India's extension system has experienced major changes since the late 1990s in governance structures, capacity, organization and management, and advisory methods. Support for public extension declined in the 1990s and donors were unwilling to fund large-scale public-sector recurrent expenditures, which led to further under financing, staffing shortages, and contraction of extension services in India (Amanor and Farrington, 1991, Rivera, Qamar, and van Crowder 2001; Birner *et al.*, 2006; Birner and Anderson 2007; Anderson 2007). Though the role of extension is increasingly being recognized world over there are criticisms about the manner in which extension work is being performed associated with different approaches of extension. Every approach has its own limitation. Kumar and Hansara (1999) listed some of the common criticism for extension in general and public sector extension in particular as bureaucracy, techno- centric work, top – down approach, providing subsidy or free inputs, emphasizing special groups and publicity stunt. Freire (1974) also mentioned that extension has the tendency to use the methods of propaganda and persuasion. Still the fact remains that the Indian public agricultural extension system is one of the largest knowledge and information dissemination institutions in the world.

Extension Systems in India

There are five major agricultural extension systems devoted to extension work in India: (i) the Ministry of Agriculture at central level, including the Indian Council of Agricultural Research (ICAR) and the Directorate of Extension (DoE); (ii) State Departments of Agriculture (DoA), as well as the State Agricultural Universities (SAUs);(iii) the Departments of Agriculture (DoA), Animal Husbandry (DAH), Horticulture (DoH) and Fisheries (DoF), as well as the Krishi Vigyan Kendra (KVKs) and, more recently, the Agricultural Technology Management Agency (ATMA) at the

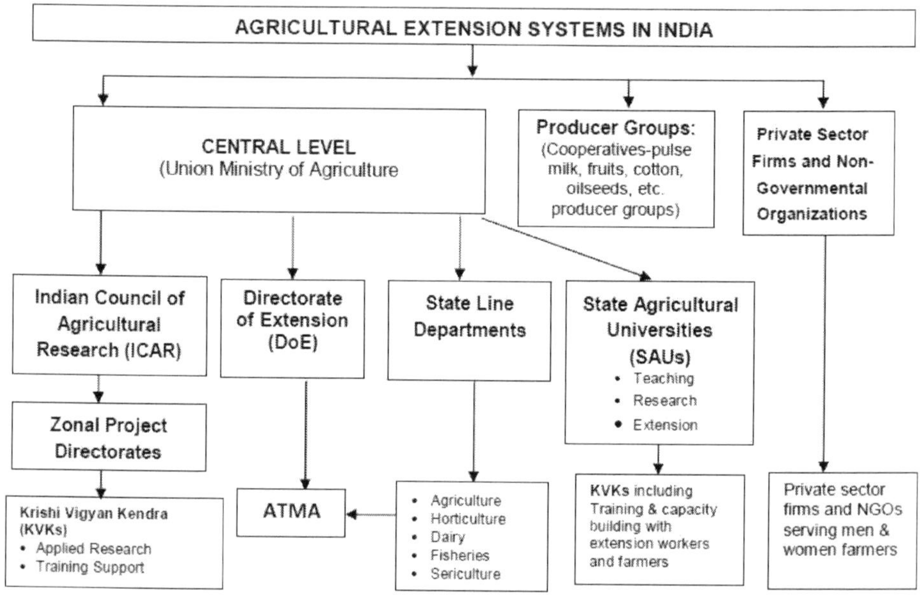

Source: Meena, M.S., K.M. Singh and B.E. Swanson

Figure 16.1: Agricultural Extension Systems in India.

District level; (iv) also, there are a wide variety of producers groups, including cooperatives and federations of milk, fruits, cotton, oilseeds, coconut, spices etc.; as well as (v) civil society organizations, such as the Non-governmental Organization (NGOs).

Public Extension System

The Public extension system played a critical role during the Green Revolution period, but in recent years, it has undergone a high level of scrutiny (Sontakki *et al.*, 2010; Pal, 2008; Joshi *et al.*, 2005). Like many other developing countries, extension services in India have been traditionally funded and delivered by the Government. In post independence era, first extension effort was Community Development Programme and National Extension Service. Later, area and target specific extension service were implemented like Small Farmers Development Agency, Drought Prone Area Programme etc. Since 1977 and early 90's most of the states have been following the Training and Visit system of extension. After the observation of unsustainable functioning of the training and visit (T and V) program by many evaluation studies (Anderson,Feder, and Ganguly2006; Moore 1984), the agricultural extension system changed its focus towards more decentralized, participatory, and demand-driven approaches (Birner *et al.*, 2006; Birner and Anderson 2007; Davis 2008; Hall *et al.*, 2000; Kokate *et al.*, 2009; Sulaiman and Hall 2008; Swanson 2009). Several efforts have been made in the public sector over the past one decade to initiate various reform measures and operational models to improve the organizational performance of this system. Yet, the challenge of enhancing relevance, efficiency, and effectiveness of the public sector agricultural extension system in meeting its organizational goals and objectives remains unresolved (WGAE, 2007; Raabe, 2008; Glendenning *et al.*, 2010; Desai *et al.*, 2011).

The demand-driven agricultural extension which still exists for more than a decade now, ushers the need for new modes of reaching out to farmers. It has potential to have significant impact on the pattern of country's farming, as it provides better opportunity to stakeholders to reflect the local information needs of farmers. There are generic problems inherent in extension functions which require collective action, and often involve government. Eight generic problems as stated by FAO (1997) are - scale and complexity, dependence of extension on the wider policy environment and other agency functions, inability to trace cause and effect, commitment and political support, accountability, liability to public service functions beyond agricultural knowledge and information transfer, operating resources and fiscal sustainability and interaction with knowledge generation. India is not alone in the world in reforming its extension and research systems. There are many countries where extension and advisory services reforms are occurring globally (Swanson and Rajalahti, 2010; World Bank, 2012).

At present there is hardly any functional linkage between farmers and marketing institution in public extension system. The present system recognizes middle men who corners major chunk of profit of prime producers as well as ultimate consumer. Also the current extension system lacks in terms of providing real time information and better export marketing platforms. To meet this challenge another reform option is the increased use of modern information and communication technologies (ICTs), but this calls for a higher level of investment in order to make them accessible by the smallholder farmers (Aker, 2011). A nation- wide survey of farmers' regarding access to extension services was done in 2003 by National Sample Survey Organization (NSSO) on 'Situation Assessment Survey of Farmers'. The survey report reveals that Sixty per cent of the farmer-households in India did not access any information on modern technologies that year. Such a huge number of farming population does not use any extension service indicates the poor organizational performance of the public extension in 2003.

It is evident that the broad objectives of decentralization and farmers' participation have been partially achieved but. However, it is reported that the reforms fall short in terms of increased accountability to farmers and being fully demand-driven. The public sector extension system has achieved only partial inclusiveness of smallholder and marginal farmers. The strategy of group approach to extension remains weak and needs strengthening at the block and village levels. It was concluded that the reform measures provide opportunities to the states in terms of flexibility, adaptability, and learning and thereby leading to the sustainability of reformed system. At the same time, huge gaps in organizational and human capacity suggest the need for long-term capacity development strategy (Suresh C. Babu,P. K. Joshi, Claire J. Glendenning, Kwadwo Asenso Okyerecan, Rasheed Sulaiman V., 2013)

Private Extension System

It is frequently reported that privatization could be a feasible solution and if privatization is appropriate, some generic problems will be positively affected. All privatization efforts report improvements in accountability, usually expressed in terms of client orientation and satisfaction. Obviously, private sector initiatives are not affected by other public service liabilities. Private-sector participation is seen as a useful means of addressing the research and extension needs of commodity-oriented farmers' and women's interest groups since it utilizes the cost advantages, capacities in research and development, and skills in multi-disciplinary and participatory research offered by that sector (Reddy and Swanson 2006). These aspects are perceived to be critical components for improving the accountability of the public sector in agricultural research and extension, for providing demand-driven support services, for encouraging the introduction of improved inputs (such as fertilizer and seed varieties), and for strengthening marketing and postharvest activities and processing.

Privatized efforts invariably claim improved efficiency, cost-effectiveness, and reduced public sector costs by servicing the needs of farmer clients who can afford to pay for the information, thus overcoming the problems of sustainability and dependence on fiscal allocations. Incentives exist for private providers of extension to maintain close links with knowledge generation agencies in order to have a marketable product. For other generic problems, the impact of privatizing extension is mixed (Swanson 2009). Over reliance on private extension risks neglect of less commercial farmers and lower-value crops. Stratification and separate, publicly-funded targeted programs are needed to counter this risk. Privatization also does not deal with the complexity of providing a socially and environmentally optimal service. Most analysts suggest a cautious, evolutionary approach toward privatization within a clearly formulated mission and strategy, along with open communication among all stakeholders. Some examples of private extension interventions are:

Mahindra Kisan Mitra (MKM), a website where the farmers can check daily mandi prices, read weather updates, latest crop advisories, and agri related news. Farmers can also read and benefit from the success stories of other fellow farmers. The Samriddhi Centers have been set up to offer farmers a wide range of agri-related services under one roof which includes insurance products, sales and service of tractors and implements, productivity demo farms, soil and water testing facilities, knowledge updates on the weather, crops, eradication of pests and diseases, mandi prices, etc., via the internet, finance facilities and agri counseling.

Tata Kisan Kendra is developed by Tata Chemicals Limited (TCL) and has played a pivotal role in giving the rural farmers access to the latest ICT tools, knowledge about modern farming technologies and information for enhanced income. In order to support the activities of TKS, which are basically franchisee outlets, the supply chain model that evolved is a 'hub and spoke' model. The hub acts as the Resource Center to cater to the needs of the TKS in its vicinity.

Hariyali Kisaan Bazaar (HKB), run by DCM Shriram Consolidated Ltd. (the fertilizer, seed and sugar conglomerate), is an example of a business that seeks to provide 'end-to-end agri-solutions' to farmers. The offer is built around a package of agri-inputs, extension, credit, and produce marketing services. Hariyali Kisan Bazaars are helping transform rural India by providing all manner of services to farmers. While the retail revolution in urban areas is going ahead at its own pace, the retailing in rural areas is also getting modernized in a unique manner to cater exclusively to the wide-ranging needs of customer-farmers.

Public Private Partnership: Paradigm Shift in Extension System

The scope of extension is increasingly becoming wider, covering all aspects of farming, from seeds to market. Farmers need information not only on best practices and technologies for crop production, but also information about post-harvest tasks including processing, marketing, storage, and handling (Van den Ban 1998, Sulaiman and Holt 2002).The agricultural development cannot be adequately addressed by public sector or private sector alone. Public investment in productivity-enhancing agricultural R and D are not sufficient in Indian scenario on the other hand private bodies are at present capable to make investments but they lack in infrastructure. It is evident from various literatures that going beyond the current linear technology transfer mode of extension requires a pragmatic and programmatic approach to the delivery of extension services. These constraints led us to realize the need and opportunities for R and D partnerships that pool assets to farmers' benefit. The public sector provides strength in infrastructure and the private organizations contribute expertise in investment strategies, knowledge expertise, marketing and delivery of products and services. So, an integration of both *i.e.* Public-Private Partnership would lead us to broaden the scope and impact of agricultural extension and to strengthen the capacity for collective action by small farmers and local institutions (SHGs, Farmer clubs etc.) to access and profitably engage in commercial markets.

In the Public-Private-Partnership model, with support of public extension system infrastructure, inputs provided by the private dealers and the farm-advisory role of agri-clinic-operators, could be made more effective by improving their capacity at the district levels. The private bodies play a leading role in providing extension services to smallholder farmers as private input suppliers would help to finance field demonstrations and training programs. The farmer-based organizations could help to fund some of the local operating costs for extension, including training and possibly stipends for community-based facilitators. Three main potential functions of PPP model could be for advancement in agriculture to meet global challenges in food security, helping in widening access to technology and adoption of the same and better linkage of farmers to markets. Agricultural Technology Management Agency (ATMA) is evolving as a prominent platform for PP-Partnerships, and has proved itself as the one that aimed at providing solutions along the entire agricultural value chain.

Agricultural Technology Management Agency (ATMA) and its Extension Approach

In the late 1990s, the Government of India (GOI) and the World Bank together pilot-tested a new, decentralized, market-driven extension model under the National Agricultural Technology Project (NATP). This new approach was designed to help farmers diversify into high-value crops and livestock enterprises, the products of which they could then sell, as a means of increasing farm income, rural employment and poverty alleviation. The Innovations in Technology Dissemination (ITD) component of NATP was implemented in 28 districts in the country. The key institution in implementing this new approach was the Agricultural Technology Management Agency (ATMA). ATMA is supposed to operate on PPPE (Public Private Partnership Extension) mode. The key elements of the ATMA model include:

(i) taking "farming-systems" approach which required the integration of extension activities across the different line departments; (ii) organizing small-scale farmers, including women, into farmer interest groups (FIGs); (iii) decentralizing extension decision-making down to the district and block levels including farmer input and. (iv) linking the farmer-interest groups to markets for the purpose of increasing farmer income and creating rural employment.

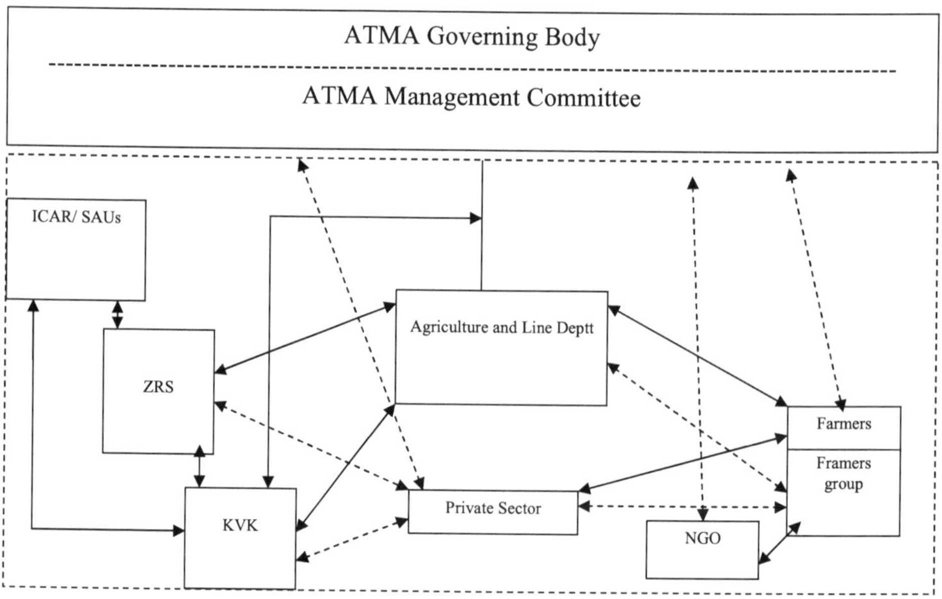

Figure 16.2: Stake Holders of Agricultural Technology Management Agencies (ATMA).

***Source*: MANAGE, Hyderabad.**

ATMA is a district level autonomous body. The extension delivery is oriented towards group approach catering to the location specific requirement of the farmers. ATMA has established a Farm Information and Advisory Centre (FIAC) at each block in the district to act its extension planning and operational arm. FIAC is supported by two groups; one, consisting of technical officers at block derived from different functional areas termed as Block Technology Team (BTT), whereas, the other is a Farmers Advisory Committee (FAC) which is a body exclusively of farmers. While BTT develops and implemnts the Block Action Plans (BAPs) in light of the Strategic Research and Extension Plan (SREP) formulated by FIAC. The FAC plays a more proactive role by scrutinizing, improving and approving BAPs, before these are referred to the ATMA Governing Body for its final approval.

As currently seen in agriculture, the private sector has became increasingly involved in meeting the many demands of the farming community. In this context, the partnership between the private sector and govt. owned extension system provides an opportunity to work together to promote extension efforts and it has emerged as one of the crucial areas in agricultural extension. A large number of ATMAs have taken initiatives to develop partnerships with the private sector in the processing industry, farmer's organizations, cooperatives, corporate bodies etc. in different areas. This partnership facilitated the dissemination of technologies, the supply of quality inputs (seed, fertilizers, micro-nutrients, bio-fertilizers, pesticides and bio-pesticides and other technological tools) and marketing of farmers' produce. For promotion of technology and market led agriculture the cases mentioned as under better describe the functioning of ATMAs:

Case I: M/s Aditya Agro-clinic and Agro Business Centre, Munger Patna-ATMA, Patna- FIGs (Market – Led Production of Basmati Rice)

Traditional rice varieties are replaced by Basmati rice as a result of partnership for production and procurement of the same. The project put ample efforts in promoting sustainability-enhancing and environment friendly technologies as well as latest improved farm practices. Some of these include integrated pest management, integrated nutrient management, intercropping, mixed cropping, green manuring, seed treatment, line sowing, drip/sprinkler irrigation, vermi compost use, zero tillage, bio-fertilizers, poly house technology, etc. A number of training programs and exposure visits for farmers were conducted through ATMA initiatives to promote these technologies/ practices. Inter-cropping/mixed cropping has been adopted on large scale in Patna and Munger. Organic farming started to take roots in Patna. (Singh and Swanson, 2005)

Case II: ATMA, Ratnagiri – KVK, Shirgaon - SAU- Dapoli - FIGs (Direct Trading of mangoes)

Alphanso is the predominant horticulture crop of Ratnagiri district. Initially growers used to sell their products at $1/4^{th}$ price through middleman. ATMA, Ratnagiri promoted 13 FIGs and trained on quality mango production, better handling through KVK- Shirgaon- SAU- Dapoli. FIGs established stalls at Mango fair, farmers got better price and consumer got cheaper mangoes, a win-win situation for both by elimination of middleman. ATMA supported to mango FIGs towards 50 per cent cost of hiring the stall national level Mango Festival organized in Mumbai.

Case III: ATMA and Safal National Exchange of India Limited (SNX), - Vellore District (Online Trading of Mangoes)

SNX would work to promote online trading in banana and mango in Vellore district through the ATMA. ATMA would arrange meetings with banana farmers in the district to introduce the tissue culture method to them and SNX would deal with trading, to enable farmers to market their produce and obtain a remunerative price. Under this partnership farmers' associations could become members of SNX and avail themselves of the on-trading service. The individual farmer had to pay a total commission of three-and-a-half per cent, of which two per cent would go to the farmers' association and the remaining to SNX. The latter also obtained a one-and-a-half per cent commission from the buyer. Once the farmers' associations became members, they could log on to the SNX website and quote their price. The SNX would contact the buyers and mediate between the two. Once the two sides agreed on a consensus price, the deal would be clinched. Quality officials of SNX would go to the fields, test the quality of the produce and take it to the nearest SNX delivery centre, from where the buyers would lift the produce after paying the amount into the SNX account through e-banking. SNX would make the payment online to the members, who, in turn, would pay through cheques to individual farmers. The objective of SNX is not to take away business from wholesale or retail traders but reduce the number of middlemen involved in the process and enable farmers to get a remunerative price. (Singh and Swanson, 2005)

ICT interventions by ATMA

The most visible impact of ICT interventions has been the increased awareness and technical capacity building through ATMA in rural areas. The ICT capacity building has also helped the ATMA staff to develop different matrices for collation and submission of monthly, quarterly, annual reports/returns, reimbursement claims, office accounting, document the project achievements and also share their successes across the state and so also at national level. ATMAs have brought out their success stories on CDs in user-friendly formats. Their reporting has improved in terms of regularity, consistency and also in terms of articulation. Another important ICT intervention, in terms of creating positive impact has been the installation of video conferencing facility at SAMETIs. The ICT initiatives under ATMA districts have promoted PPP in various models. Yet another ICT initiative of ATMA is promoting the set-up of 'Information kiosks' in private sector and providing them the technical guidance. These information kiosks are promoting ATMA philosophy and creating awareness among the farmers' groups. Such ICT interventions have made significant impact in improving Research-Extension-Farmers linkages by providing reliable connectivity and state-of-the-art infrastructure. It also provides in-depth and on the job Training support to ensure optimal use of ICT infrastructure. The major aim is to promote media linkage and networking of various of agricultural institutions and stakeholders through the use of e-mail, custom-made software and other ICT tools. The sustainability aspect is also taken care of by providing content generation capacity within the ATMA system for all the ICT infrastructure and increasing number of ICT applications (Singh and Swanson, 2005).

Future Prospects

The inclusiveness of extension services remains a major concern. Considering the prevalence of smallholders in Indian agriculture and the complexity of the problems confronting them, suitable extension strategies need to be formulated. The growth of smallholder agriculture will be determined by the extent to which institutions of research and extension are attuned to their priorities a general weakness observed in agricultural extension is the low adoption of new agricultural technologies by poor and small farmers. As the present public extension systems are struggling to address the issues and concerns of small and poor farmers. All India value of extension effectiveness index is 47 per cent indicating moderately effective extension mechanism in India. Hence a search for new extension models that are more effective, efficient, and responsive to different categories of farmers is felt essential. However, the major issue related to the alternative extension model *i.e.* Public Private Partnership Extension (PPPE) is, though the technology is transferred with free of cost, only 25-30 percent technologies are reaching the farmer's fields. What would be the fate of remaining percent of technologies, if extension services are privatized?

One important function of extension is technology transfer to the farmers and getting feedback. If the services are privatized, extension personnel's interaction with farmers' will be restricted and access to free flow of information with the agricultural knowledge system might be hampered. However, the modern and information rich extension system requires greater integration of different key stakeholders through inclusive-models like that of PPP at the multi levels of implementation. At the same time, the extension staff at those levels needs systematic training in areas like planning, resource management, monitoring, evaluation and documentation. Also, the incorporation of ICT requires capacity building and infrastructure support. Hence appropriate linkages between public and private extension systems needed to be developed and has to be implemented in phased manner wherever it is applicable through rational and judicious blending of public and private services. A healthy win-win atmosphere with mutual trust has to be created among the individuals and organizations for providing services. Policy makers and practitioners have to consider the global issues of economic liberalization, food security and sustainability when framing policies for PPP, particularly with reference to agricultural extension services targeting resource poor farmers.

References

Aker, J.C. 2011. Dial "A" for agriculture: A review of information and communication technologies foragricultural extension in developing countries, Agricultural Economics, 42: 6.

Amanor, Kojo and John Farrington. 1991. NGOs and Agricultural Technology Development. In Agricultural Extension: Worldwide Institutional Innovation and Forces for Change (W. M. Rivera and D. J. Gustafson, eds.). Amsterdam: Elsevier.

Anderson, J.R., Feder, G. and Ganguly, S. 2006. The Rise and Fall of Training and Visit Extension: An Asian Minidrama with an African Epilogue. Washington, D.C.: World Bank.

Anderson, J. R. 2007. Agricultural advisory services. Background paper for the World Development Report 2008. Agriculture and Rural Development Department. Washington, D.C.: World Bank.

Babu, Suresh C. and Joshi, P.K. and Glendenning, Claire J. and Kwadwo, Asenso-Okyere and Rasheed, Sulaiman V., 2013. "The State of Agricultural Extension Reforms in India: Strategic Priorities and Policy Options," Agricultural Economics

Research Review, Agricultural Economics Research Association (India), vol. 26(2).

Birner, R. *et al.*, 2006. From best practice to best fit: a framework for analysing agricultural advisory services worldwide. Washington, DC: IFPRI.

Birner, R., K.Davis, J.Pender, E.Nkonya, P.Anandayasekeram, J.Ekboir, A.Mbabu, D.Spielman, D.Horna, S.Benin, M.Cohn. 2006. From "best practice" to "best fit": A framework for analyzing pluralistic agricultural advisory services worldwide. DSGD Discussion Paper 37. Washington, D.C.: International Food Policy Research Institute

Birner, R., and J. R. Anderson. 2007. How to make agricultural extension demand-driven? The case of India's agricultural extension. Discussion Paper 729. Washington, D.C.: International Food Policy Research Institute.

Davis, K. 2008. Extension in sub-Saharan Africa: Overview and assessment of past and current models, and future prospects. Journal of International Agricultural and Extension Education 15(3): 15–28.

Desai, B., D'Souza, E., Mellor, J.W., Sharma, V.P. and Tamboli, P. 2011. Agricultural policy strategy, instruments and implementation: A review and the road ahead. Economic and Political Weekly, XLVI.53: 42-50.

FAO (Food and Agriculture Organization of the United Nations). 1997. Improving Agricultural Extension: A Reference Manual. Rome: FAO.

Freire, P. 1974. Education for Critical Consciousness. London. Sheed and Ward.

Glendenning, C.J., Babu, S.C. and Asenso-Okyere, K. 2010. Review of the Agricultural Extension in India, IFPRI Discussion Paper 01048, International Food Policy Research Institute. Washington, D.C.

Hall, A., *et al.*, 2000. Coping with new policy agendas for agricultural research: The role of institutional innovations. ICAR Policy Brief 13. New Delhi: Indian Council of Agricultural Research.

Joshi, P.K., Pal, S., Birthal, P.S. and Bantilan, M.C.S. 2005. Impact of Agricultural Research: Post Green Revolution Evidence from India, NCAP/ICRISAT, New Delhi.

Kumar, B. and Hansara, B.S.1999. Extension Education for Human Resource Development. Concept Publishing Company, New Delhi.

Kokate, K. D., *et al.*, 2009. Farmers-led extension: Experiences and road ahead. Indian Research Journal of Extension Education 9(2):18–21.

Moore, M. 1984. Institutional development, the World Bank, and India's new agricultural extension programme. Journal of Development Studies 20(4): 303–317

Pal, S. 2008. Agricultural R and D policy and institutional reforms: Learning form the experiences of India and China, Economic and Political Weekly, pp. 145-155.

Raabe, K. 2008. Reforming the Agricultural Extension System in India. What do We Know about What Works Where and Why? IFPRI Discussion Paper 00775, Washington DC.

Reddy, M.N. and Swanson, B. 2006. Strategy for up-scaling the ATMA model in India, In: Proceedings of the Association for International Agricultural and Extension Education, Ed: J.R. Vreyens. AIAEE 22nd Annual Conference, Clearwater Beach, Florida, USA.

Rivera, W. M., K. M. Qamar, and L. V. Crowder. 2001. Agricultural and rural extension worldwide: Options for institutional reform in developing countries. The Food and Agriculture Organization, Rome.

Singh, J.P., Singh. Krishna M. and Swanson. Burton E. 2005.A good practice note on ATMA model in India. Thematic Group Sustainable Agriculture Systems, Knowledge and Institutions (SASKI).World Bank.

Sontakki, B.S., Samanta, R.K. and Joshi, P.K. 2010. Redesigning Agricultural Extension in India: Challenges and Opportunities, Summary proceedings and recommendations of NAARM-IFPRI workshop held during 20-21 August 2010. National Academy of Agricultural Research Management, Hyderabad, Andhra Pradesh.

Sulaiman, R. and Holt, G. 2002, Extension, Poverty and Vulnerability in India: Country Study for the Neuchatel Initiative, Overseas Development Institute, London.

Sulaiman, R., and A. Hall. 2008. The fallacy of universal solutions in extension: Is ATMA the new T and V? <www.innovationstudies.org/index.php?option=com_myblog and show=The-Fallacy-of-Universal-Solutionsin-Extension-139.html and Itemid>. Accessed October 4, 2014.

Swanson, B., K. M. Singh, and M. N. Reddy. 2008. A decentralized, participatory, market-driven extension system: The ATMA model in India. Paper presented at Advancing Agriculture in Developing Countries through Knowledge and Innovation, Addis Ababa, Ethiopia.

Swanson, B. and Rajalahti, R. 2010. Strengthening Agricultural Extension and Advisory Systems: Procedures for Assessing, Transforming, and Evaluating Extension Systems, Agricultural and Rural Development, Discussion Paper 45, The World Bank, Washington DC.

Van den Ban, A.W. and Hawkins, H.S. 1998.Agricultural Extension. New Delhi. CBS Publishers and Distributors.

WGAE (Working Group on Agricultural Extension) 2007. Recommendations of Working Group on Agricultural Extension for Formulation of Eleventh Five-Year Plan (2007–12), Planning Commission, Government of India, New Delhi

World Bank. 1997. Rural Development: From Vision to Action. Sector Strategy, Environmentally and Socially Sustainable Development Studies and Monographs Series 12. Washington, D.C.

World Bank. 2007. World development report 2008: agriculture for development. Washington, DC, The World Bank.

World Bank, 2008. World Development Report 2008: Agriculture for Development, The World Bank, Washington, D.C.

World Bank. 2012. Agricultural Innovations Systems – An Investment Source Book, The World Bank, Washington D.C. 658 p.

Chapter 17
Gender Mainstreaming in Agriculture and Allied Sectors

***Shilpi Verma*[1] *and Shobhana Gupta*[2]**

[1]*RVSKVV–Krishi Vigyan Kendra, Neemuch, M.P.*
[2]*Directorate of Extension Services, RVSKVV, Gwalior, M.P.*

Nothing changes the gender equation more significantly than women's economic freedom.

— Gloria Steinem

Women are key to the development challenge. Throughout the developing world, women are at a disadvantage at the household, community, and societal levels. Within the household, women have less access to and control over resources and limited influence over household decisions. Beyond the household, women have limited access to communal resources, are under-represented in public decision-making bodies; have limited bargaining power in markets (such as the labor market), and often lack opportunities to improve their socioeconomic position. Therefore, efforts to reduce gender inequality are required on multiple fronts. Over the last several decades a number of strategies have emerged and evolved to promote gender equity in development efforts. Yet debates regarding the relative efficacy of these strategies remain.

Gender mainstreaming has been associated with more failures than successes. While there have been some positive gains for gender equity in the ten years, a number of factors—including the challenging policy environment within which gender mainstreaming processes operate, inadequate resources allocated to this work, institutional features that have blocked change, and the way in which gender mainstreaming processes have been implemented—contributed to the overall failure of gender mainstreaming. While advocates of gender mainstreaming envisioned both institutional and social transformation, in practice, bureaucracies have not proven to be effective agents of social transformation. Moving forward, should involve strengthening the capacity of states and development bureaucracies to deliver on their own operational mandates and developing realistic strategies and workable alliances in light of the constrained institutional environment.

Grounded in feminist theoretical frameworks and intended to make mainstream institutions agents of social change, gender mainstreaming refers to a wide set of strategies and processes.

The Economic and Social Council of the United Nations (ECOSOC) defines gender mainstreaming as "...the process of accessing the implications for women and men of any planned action, including legislation, policies or programs in any areas and at all levels. The strategy of making women's as well as men's concerns and experiences an integral dimension of the design, implementation, monitoring and evaluation of the policies and programs in all political, economical and societal spheres, so that men and women, in fact, benefit equally; but that the goal of gender mainstreaming is gender equality." Yet gender mainstreaming is neither a clear agenda for institutional transformation nor a clear agenda for gender transformation and social change. Thus, in practice, gender mainstreaming has often involved adopting a gender policy, creating a gender unit to work on organizational programs, mandatory gender training, and increasing the number of women staff and managers. In the worst cases, gender mainstreaming has been used to stop funding for women's empowerment work, and to dismantle many of the institutional mechanisms such as women's units and advisors created to promote women in development in the name of integration. This shows, that the transformational change expected to result from gender mainstreaming has not occurred.

In 2010, the Food and Agriculture Organization of the United Nations (FAO) famously concluded "if women had the same access to productive resources as men, they could increase yields on their farms by 20–30 percent. This could raise total agricultural output in developing countries by 2.5–4 percent, which could in turn reduce the number of hungry people in the world by 12–17 percent." Whatever the validity of this calculation, the reality is that food security today depends, even more than in the past, on combating discrimination against women, in order to allow women and female-headed farming households to produce under better conditions.[2]

The Eleventh Five Year Plan visualizes "faster *and more inclusive growth*". This implies that we need to/for:

- ✰ Shift the plan priorities towards the social sectors
- ✰ Put appropriate institutions and linkages in place to bring good to the common man
- ✰ The States to play a facilitating role in encouraging individual-institutional linkages

Why Women in Agriculture

- ☆ Eradicating poverty
- ☆ Ensuring food security
- ☆ Promoting their own well being
- ☆ Their increasing stake in agriculture
- ☆ Decreasing economic contribution
- ☆ Ownership in livestock, land, etc. is limited

Statistics

While only 53 per cent of the male population is involved in agriculture, the corresponding numbers for the female population was a 85 per cent of all rural female workers. According to the 2011 census, women account for 586.47 million in absolute numbers and the round of the National Sample Survey Organization's (NSSO) data, women accounted for 38.51 percent of the total rural workers engaged in primary sector during 2007-08[1].They constitute a significantly large proportion (about 80-83 percent) of the women workers as compared to 63-66 percent among rural male workers. Women also play a significant role in the various allied activities - as per the 'time use survey' covering six states in the country during 2000, women spent almost an equal amount of time on crop cultivation and allied activities. Unfortunately they receive payment for only 60 per cent of the time they spend in agriculture for a substantial part of their work is likely to be on family farms.

Context of Agrarian Crisis

- ☆ Low size of landholding
- ☆ Faulty policies and structures
- ☆ Lack of technology
- ☆ Sale of spurious seeds and pesticides
- ☆ Debt at very high interest rates
- ☆ No health insurance and crop insurance
- ☆ NCF agrees that there is no agricultural policy that supports institutional arrangements

Solutions

- ☆ Firstly, the focus has to be on sustainability
- ☆ Find ways to encourage farmers towards food production over commercial crops
- ☆ Strengthen capital related investment in agriculture and rural infrastructure

Invisibility of Women Work

- ☆ Conceptual Biases in Measurement
  - Non recognition of Women's role in decision making

- Historical and Complex Causes reinforced by social, cultural, political and religious practices and beliefs

☆ Serious inaccuracies and measurement failure in the recording of the work that women do due to conceptual and operational (enumerators' and respondents') biases at the time of data collection.

☆ Engendering Agriculture reduced to women's participation in training programs for technology dissemination or micro credit.

☆ Non involvement of women in institutional work

Recommendation

☆ Correct the statistical invisibility of women's work through preparation of an account that should include in detail the work that women undertake.

☆ Policies and funds allocation need to take cognizance of this, and address women's needs.

Adequate attention is given to the educational process through which women engage with the institutions in an informed and empowered way. Agricultural education be made gender sensitive and research, development, extension and services be engendered to give due recognition to the multiple role played by women agriculturists.

Evolution of Policy

Gender mainstreaming started from the VI Five Year Plan when *'opportunities for independent employment and income'* for women was recognized as a necessary condition for raising social status of women.

Plan	*Focus*
6th Five Year Plan	Shift from Welfare to Developmental Issues
7th Five Year Plan	Raising Economic and Social Status of Women
8th Five Year Plan	Increased Emphasis on Economic Activities
9th Five Year Plan	From Development to Empowerment
10th Five Year Plan	From Women Alone to Gender Mainstreaming
11th Five Year plan	Propose to Move Towards a Holistic Approach

Persistence of Partial/Compartmental Approach

☆ Three components of Gender Mainstreaming Approach.

- Women's Empowerment
- Capacity Building
- Access to Inputs as well as technology and resources

☆ Need for a coordinated approach across Ministries

☆ Schemes undertaken by the various Ministries will ultimately converge towards the goal of women's empowerment.

Introduction of Component Plan for Women in all development plans during the Ninth Five Year Plan was a step in this direction

Tasks for Gender Mainstreaming in Agriculture	*Main Ministries*	*Role of Ministry of Agriculture*
Women's Empowerment (human capital formation, exposure, leadership, autonomy, Self esteem, and food security)	MoA, MoRD, Social welfare, HRD, Health	Gender Focused Strategy for Agri. Growth (main contributor along with other Ministries)
Capacity Building in Agriculture (dissemination of information and technology)	MoA	Various Extension and Training Programmes (Almost the sole contributor)
Access to Agricultural Inputs (including land, water and credit besides agri-inputs)	MoRD, MoA, MoEF	Access to Agricultural Inputs, Formation of SHGs, Marketing Facilities (partial contributor with MoRD and MoEF having a major control over property rights regimes)

In a country where majority of the population still depends on the primary sector for their livelihood, gender inequalities (social, economic and political) can be challenged by granting ownership and control over land to women. It can be done through:

a) Inheritance

b) Direct government transfers

c) Purchase or lease from the market

To enhance women's land access from these three sources, a range of initiatives are needed, including joint land titles in all government land transfers, credit support to poor women to purchase or lease land from the market, increase in legal awareness and legal support for women's inheritance rights, supportive government schemes and recording of women's inheritance shares, and so on. There is also need for reliable, fair and accessible mechanisms such as social audit with greater participation of women in the audit bodies for resolving disputes and providing remedies in matters related to tenure and security of lease.

Approach for Gender Mainstreaming: Needs

☆ The present approach for gender mainstreaming needs strengthening in terms of:

- Adoption of a more holistic and comprehensive approach to impact multiple activities and requirements of individual farming women.
- Effective implementation perhaps, in partnership with NGOs.

All these would involve fresh thinking and an overhauling of the design, resource mobilization, and implementation machinery.

Recommendations for Gender Mainstreaming

- ☆ For the Plan to be Gender and Poverty Sensitive it must allocate funds so that first priority is given to ensuring food security and access to food for all.
- ☆ Establish an independent regulatory authority which should be required to regulate, test and certify quality of inputs *e.g.* seeds, pesticides.
- ☆ Encourage horticultural activities, nursery raising, nursery maintenance, hybrid seed production, and tissue culture propagation of fruits and flowers as these are remunerative employment options for women. The fruit and vegetable processing industry also has high employment potential.
- ☆ Allocation funds to enable extension of the NREGA such that women in each household gets access to at least 100 days work in each year and their right to get work does not get subsumed within the household.

Some Attempts and Initiatives taken by NGOs

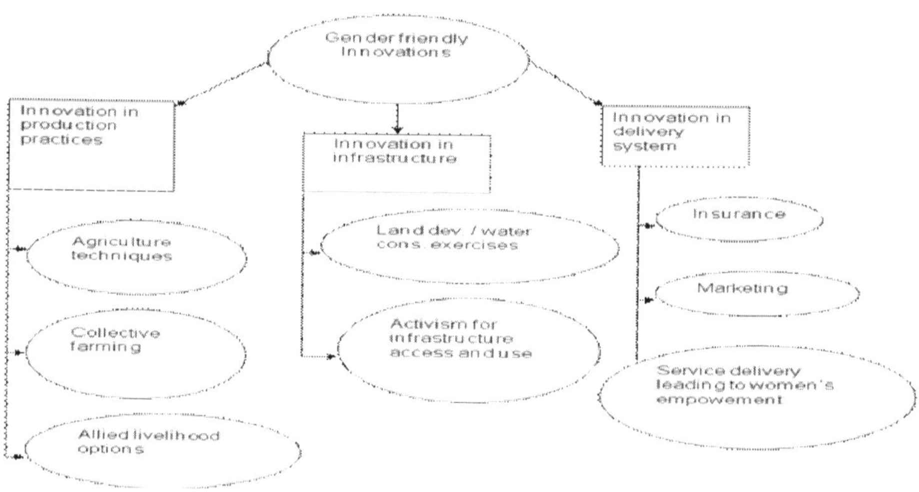

Technological Incorporation in Agriculture

Since the time of Green Revolution, incorporation of technology in agriculture sector was promoted by the government. Behind it, there was a key notion - technology reduces drudgery, increases productivity and leads to a better and healthy life. Although within a vast country like India where topographic conditions differs a lot from region to region, it was not easy to boost the agriculture sector on technological front. Where earlier it was mainly men's control over these innovations, after many years access to agricultural technology by women has been recognized by the Government and 12th FYP (Five Year Plan) mentioned that:

(12th Five Year Plan)

Technology transfer to women would be prioritized in all aspects of farming and farm management, including dry land farming technologies, animal husbandry, forestry, sustainable natural resource management, enterprise development, financial management and leadership development. They would be provided training in pre and post-harvest technologies. To train women farmers in new technologies and practices, gain access to information on schemes and subsidies, training in crop planning and so on. Special Resource Centres would be provided. Women and young girls will be given training in the use and repair of bore wells with special focus on promoting low cost irrigation.

Although, small land holding pattern and existing power relationship in village settings are some major challenges which raise serious concerns over the affordability and accessibility for such technologies. Technological inputs and incorporation of global market changed the agricultural practices to certain extent in the past but it also put the farmers in a new domain where their lack of expertise makes them more vulnerable. This vulnerability severely affects women while putting their livelihood practices at severe risks (*i.e.* market, environmental and social). (Swain and Singh, 2013)

Innovation in Agriculture Practice

Agriculture Techniques

Anar De, (Gujarat) has done a very successful and remunerative program in cultivation of mangoes and cashew nuts. Intercropping, vermicomposting for women were demonstrated. Mushroom cultivation and Tissue culture were also successful.

Collective Farming

Sewa organized the landless women agriculture workers into a co-operative to cultivate wasteland. Acquiring the land, systematical planning and environment friendly agriculture practices, including horticulture, agro-forestry and rainwater harvesting.

Activity of nursery raising is now accepted as an alternative source of employment for agriculture workers. To minimize the costs, some *Mahila mandals* have started collecting their own seeds.

Allied Livelihood Options

- *Cooperative Development Foundation* in Andhra Pradesh has rooted the cooperative movement in the rural people in the districts of Warangal, Medak, and Karimnagar of the Telangana Region.
- This started with farmer's cooperatives for credit, marketing and inputs. The women's cooperatives extended it to provide credit to milk and related products. This worked very successfully to improve their livelihoods
- *UNDP*, under its women in agriculture projects, undertook 21 districts spread across Andhra Pradesh, Uttar Pradesh and Orissa.
- Provided resources and opportunities to women's groups to increase and improve their private agricultural land, experiment with a range of farm

and off-farm based economic activities, and improve their access to sustainable farm practices, quality seed and inputs.

- ✰ Support was also provided for building rain water harvesting structures, de-silting tanks/ponds and reviving irrigation systems to improve the productivity of the land.
- ✰ *Core issues for women in agriculture* projects are access to land and working on farms collectively. Hence access to productive assets including land, credit, technologies, inputs and subsidies were made available to women's groups.

Innovations in Infrastructure

Land Development/Water Conservation Exercises

SEWA organized the women agriculture workers/farmers into Sabarkantha Women Farmer's Association which initiated watershed development to check soil erosion.The cooperative also works to organize the women into their own SHGs and provides the necessary training for leadership development, awareness generation, and capacity building.

Activism for Infrastructure Access and Use

Dharangrast Parishad working with *dam project affected people* in South Maharashtra came up with some innovative demands which show the concern for women:

Women would have an equal access to the water

Lands are given in joint names of the man and the woman of the household.

MASUM from Pune district has launched a successful campaign to ensure joint pattas for women and men and land-rights of female headed households.

Innovations in Delivery System

Insurance Sector

Swayam Shiksha Prayog has started a community based health insurance plan, *Sakhi*. Program benefits include reimbursements for hospital expenses of Rs. 5,000, community-level outpatient delivery (OPD) services, discounted rates and various health education workshops and programs.

Marketing

Economic associations promoted by *SEWA* in different districts have come together and formed their own *women's marketing network—Sewa Gram Mahila Haat (SGMH)*.

The essence is to strengthen the rural economy through rural procurement and rural distribution. The focused areas are agriculture, salt, gum and handloom and handicraft.

External and Internal Environment

Reviewing and accessing the availability and utility of gender-disaggregated data on women in agriculture and measures for effective generation of needed data, needed for policy making in accordance with the prevailing scenario is very important:

Issues

The data system capturing women's work simultaneously reflects on issues like:

- ☆ What comes out of the work carried out by women and men;
- ☆ Who decides the division of work;
- ☆ What are the conditions within which the work takes place; how are the benefits shared and controlled;
- ☆ What is the perceived notion of autonomy;
- ☆ To what extent does poverty (or economic well being) impinges on women's work burden;

Another Picture (*Source*: The Hindu Business line, 30 December, 2005)

- ☆ Advisor to the Chennai-based M. S. Swaminathan Research Foundation (MSSRF), on education, communication and gender, *Mrs Mina Swaminathan* feels that"people in various areas of agriculture — the extension departments, the Ministry, the agricultural universities, the ICAR or the research system, the Krishi Vigyan Kendras (KVK). the entire agricultural bureaucracy. is gender insensitive.
- ☆ They're supposed to be serving the farming community but they see only one half of this community — the men. The entire agricultural establishment does little for women because they're not aware of women and their problems.
- ☆ Historically the academic, research and other streams of agriculture have been headed by men "and the traditional perception of the kisan is male.
- ☆ Another drawback on the gender front in agriculture is whenever a poor women is considered for betterment we immediately talk about starting SHGs
- ☆ In horticulture or dairy, where labour is provided usually by women, they do not get the profits women take care of the cattle, milk the cows, etc, but in the milk co-operatives, the members are mostly men. Right from the early days the women do all the work in the milk co-operatives, but the profits come in the name of men.

The way forward, should involve strengthening the capacity of states and development bureaucracies to deliver on their own operational mandates and developing realistic strategies and workable alliances in light of the constrained institutional environment. In addition, the range of diverse strategies all termed gender mainstreaming—including policy reform, advocacy, capacity building, analytical frameworks, program development, and monitoring systems—should be disaggregated and analyzed in terms of their particular gains and failures. This would facilitate strategic thinking about what particular institutions are well positioned to accomplish, and what they can be held accountable for. Demand for change should also come from civil society, including women's organizations and networks. This would hold institutions accountable and help to ensure that their commitments are achieved.

Velayudhan (2009) also felt the same and mentioned that land entitlement by women is important in the context of growing feminization of agriculture. Supporting women farmers would also enlarge the information base of farming because in many regions women know/practice more than men about indigenous seed selection and cultivation methods.

Gender in Government Policies and Plans

Having analysed the scale of the challenges and its multi-faceted nature and deep-rooted patriarchy that is prevalent, it is important to recognize the evolution in policy approaches towards gender in agriculture since the Sixth Plan, when 'opportunities for independent employment and income' for women was recognized as a necessary condition for raising the social status of women. The present approach of `gender mainstreaming', which means that women have to be part of all the schemes/programmes of the agriculture sector and the strategy of agenda setting aims to provide structural, legislative, and material resources so that women can participate and benefit on par with male farmers by setting their own agenda. According to the National Agricultural Policy, there are three components of Gender Mainstreaming Approach. These are: Women's Empowerment, Capacity Building, and Access to Inputs as well as technology and resources.

Since many of the programmes most relevant for women are implemented at the third tier of governance - Panchayati Raj Institutions —success in achieving the outcomes depends critically on women's participation in governance and their empowerment with respect to programme implementation. While realizing the key elements for Gender Equity, the Twelfth Five Year Plan addressed gender issues within the following categories:

1. Economic Empowerment
2. Social and Physical Infrastructure
3. Enabling Legislations
4. Women's Participation in Governance
5. Inclusiveness of all categories of vulnerable women
6. Engendering National Policies/Programmes
7. Mainstreaming gender through Gender Budgeting

Putting small farmers and women farmers at the heart of 12th plan needs to be applauded. As the movement at the national and various state levels towards women's land entitlement including enactment of legislations is laudatory (Rajasthan, Odisha, Chhatisgarh, Kerala, West Bengal, Assam), so are positive/progressive legislations like the tabling of the National Food Security Bill after its passing by the Parliamentary Standing Committee, the emphasis on PDS reforms and enactment of a very progressive, almost universal, Chhattisgarh Food Security Act.

Making the Vision a Reality

The government's endeavors to increase women's employability in the formal sector as well as their asset base are evident from the 12th FYP document. The government's commitment to improve the conditions of self employed women, with special focus on women's workforce participation, particularly in secondary and tertiary sectors, however, does raise serious concerns over the future implications for women's status in primary sector. Plan document ensured decent work for women by their financial inclusion in agriculture and manufacturing while extending land and property rights. But, it is also important to note that the major impediments mentioned in the document which affect women's participation as the workforce, particularly in secondary and tertiary sectors, is the lack of skills. Abundant focus on skill development puts a scanty picture of State's willingness to improve women's stature in agriculture sector.

The government of India can achieve a lot when the public discourse and political will are at a tipping point. The concrete agenda for the next five years (not in the order of importance or chronology) should include the following but even push the ambitions further, *i.e.*

1. Generating gender disaggregated data for reportage, monitoring and programming
2. Supporting the policy and programme implementation environment for women's collectives
3. Engendering the department from highest policy making level down to the extension services
4. Operationalising a coordination mechanism between Women and Child Development Department and the Agriculture Department for linking agriculture, nutrition and recognizing spousal violence's impact on malnutrition
5. Supporting research towards drudgery reduction
6. Generating evidence and policy on farmgate discrimination

It is only a maximalist vision and sincere resourcing (human and financial resources) coupled with prioritization of agricultural sector that can bring meaning to the platitudes in the 12th Five Year Plan document.

References

Agarwal, B. (1997); A Field of One's Own: Some Salient Features of the Book; In Rao, N. and Rurup, L. (Ed.) *A Just Right: Women's Ownership of Natural Resources and Livelihood Security*; Friedrich Ebert Stiftung; New Delhi

Ali, J. (2007): Livestock sector development and implications for rural poverty alleviation in India. *Livestock Research for Rural Development. Volume 19, Article* #27.http://www.lrrd.org/lrrd19/2/ali19027.htm; Accessed January 6, 2011

Chavan, P. (2008); Gender Inequality in Banking Services, Economic and Political Weekly, Nov 22

De Schutter, Olivier (2012). The Recivilisation of Men by Women, Oxfam Discussion on "Making the Food System Work for Women", November 2012

Ghosh, Jayati (2012). Nutrition Policies that work for Women, Oxfam Discussion on "Making the Food System Work for Women", November 2012

GOI (2012); 12th Five Year Plan 2012-17, Planning Commission, New Delhi

International Institute of Population Sciences, National Family Health Survey III (2005-6), 2007

International Institute of Population Sciences, Gender, Equality and Women's Empowerment in India, National Family Health Survey III (2005-6), 2007

Kabir, Rokeya, Working Harder isn't working, Oxfam Discussion on "Future of Agriculture", December 2012

NCEUS (2008); A Special Programme for Marginal and Small Farmers; National Commission for Enterprises in the Unorganised Sector; New Delhi; December 2008

Saxena, N.C. (2012), Hunger and malnutrition in India, IDS Bulletin, Volume 43, Special Issue 1, July 2012

Shabodien, Fatima (2012). Women Farm Workers Dying for Food, Oxfam Discussion on "Making the Food System Work for Women", November 2012

Shah, Amita (2012). Priority Changes for Strengthening Women's role as Producers, Processers and Providers of Food and Nutrition, IDS Bulletin, Volume 43, Special Issue 1, July 2012

Sulaiman, R.V., Jafry T., and Ashok M.S. (2003); Cafeteria for Women in Agriculture, NCAP Working Paper 4, National Centre for Agricultural Economics and Policy Research, New Delhi

Swain, Biraj and Singh, Ranvir (2013). Women in Agriculture: Recognizing them and Re-prioritizing them? Ecological Democracy, Vol.1 Issue 3 September 2013

Usami, Yoshifumi (2009). A Note on Recent Trends in Wage Rates in Rural India, Foundation of Agrarian Studies, 2009

Velayudhan, M. (2009); Women's Land Rights in South Asia: Struggles and Diverse Contexts; Review of Women's Studies; Economic and Political Weekly, Volume XLIV No. 44, October 16

Chapter 18

Strategic Role of KVKs for Empowering Farmers through Market-Led Extension

S.R.K. Singh, Anupam Mishra and Prem Chand

ICAR-Agricultural Technology Application Research Institute, Jabalpur, M.P.

Marketing of the fresh produces at remunerative prices is a tedious process in rural areas due to lack of nearby market and difficult transportation. Most of the farmers sell their produce either to the contractors/middle men or within their locality. In such situation, KVK can promote marketing role to help them in getting the actual price of their produce. KVK through market-led extension can help the farmers by advising them on the product planning, selection of the varieties, selection of the market outlets, and preparation of the products for the markets. The farmers could be trained in maintenance of quality specifications and creation of the quality testing facilities within the reach of the farmers. Training and teaching aids in the field of agricultural marketing should be developed in regional languages, also. Further, farmers could be trained in the area of marketing cost reduction through direct marketing by the farmers and cooperative or group marketing. KVK scientists through the development agents can pursue the buyers to purchase the fresh produces from fruit and vegetable belts This will strengthen the market linkage of the farming community and make them experienced with new learning of the marketing process. Moreover, KVKs could focus on the quality parameters through grading; packaging, standardization, etc. favourable to enhance their profits. Therefore, it is important for KVKs to emphasize to form cooperative societies/Farmer Producer Organization (FPO) for purchasing improved inputs, and sell their products to the desired markets. Extension strategies especially for marketing could encourage them in formation of such farmers' organizations so that the overall socio-economic conditions could be improvised through getting higher income and employment. In all these processes, KVK can play a vital role in educating and training of farmers to reap the real benefits of the better technological interventions resulting into the higher production lead to the higher income.

Introduction

Marketing is an important issue in the today agriculture particularly for the fresh produce. The farmers are well equipped with the production and protection technologies and wish to get advice on marketing aspects for raising their profit from their farming system. In such situation, market-led extension is in dire need for the rainfed farmers. In the liberalized economy, farmers become more market-oriented, so KVK scientists and extension workers need to be in a position to advise them not only how to grow various crops but also on how to successfully market them. For this, a comprehensive knowledge of produce handling, storage and packaging is essential. Extension workers has to develop some extension techniques especially for marketing of fresh produce so that farmers can harness maximum benefits both from higher production and better price of their farm produce. Moreover, it is very important in Madhya Pradesh, as the state has great potential in producing fruits and vegetables throughout the year due to its climatic and edaphic conditions. This paper aims to highlight the important aspects of the market-led extension especially for fruits and vegetables. And this will be very useful for extension workers, change agents, NGOs personnel in guiding the farmers on the described aspects.

A. Marketing and Market-Led Extension

Marketing

The 'marketing' of fresh produce includes the entire process of directing the flow of goods and services from producer to consumer. This process includes not only the commercial transactions of buying and selling, but also the physical distribution aspects. For fresh produce, the physical distribution itself includes all the handling and movement features of harvest, loading and unloading, grading, packaging, storage and subsequent dispersal through the markets to the consumers. In addition, marketing as an entire process also includes the vital auxiliary functions of production planning, production and dissemination of market information, financing of markets and their administration, the activities of marketing intermediaries, the provision of training and extension to individuals and groups involved in marketing, and research activities which seek to improve the marketing system in some way.

Market-led Extension

It includes the marketing process judicious blended with suitable extension techniques for educating and facilitating the farmers in the area of effective marketing. It emphasizes on the training of the fruits and vegetable growers in the harvest time, post-harvest management, and suitable market information. For example- If a small farmer and member of a cooperative can grow capsicum for sell to different metros through a regional agency with various buyers with different variety, grade and

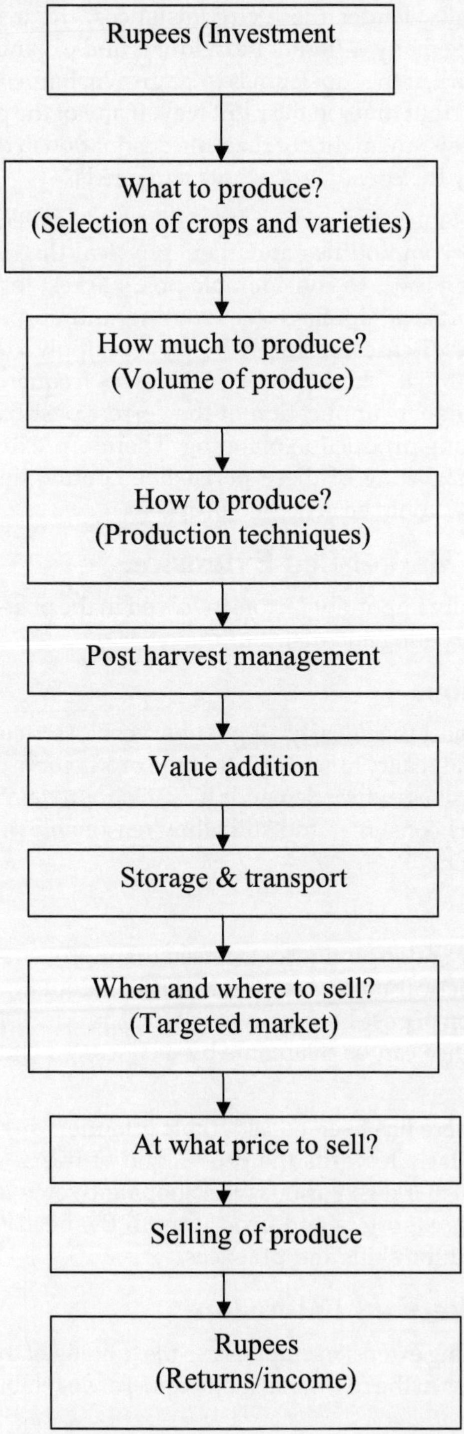

Figure 18.1: Money-to-Money Concept of Market-Led Extension.

packaging requirements. Under these circumstances, the marketing of a single commodity can involve many different individuals and organizations, with strong communication network, if the capsicum is to be grown, harvested, graded, packed and transported at the right times in the right way. If any of the people concerned do not do their job the capsicum might not reach its destination in the right condition to make a sale and satisfy the consumers' wants and needs.

Fresh fruits, vegetables and root crops, collectively called fresh produce, are inherently perishable commodities and their physical distribution through the marketing system often leads to considerable post-harvest losses through quality downgrading, physiological spoilage (ea. rooting and sprouting), pathological spoilage from pests and diseases, or simply from over-supply to a market that cannot absorb the volume sent to it. These post-harvest losses frequently occur because of inappropriate or inadequate application of the correct post-harvest practices and considerations, including production planning. Therefore, extension interventions are required in the marketing of these perishable commodities so that both the producer and consumer could be benefited properly.

B. The Need for Market-Led Extension

There are two vitally important factors involved in the post-harvest loss of fresh produce during its marketing and distribution:

Price Considerations

If cost were no object then nearly all post-harvest losses could theoretically be controlled. The practical reality however is that the production, distribution and sale of fresh produce is a business which demands that any improvements in the marketing must be paid for by the consumer and still allow reasonable profit margins for the growers and distributors.

Ignorance

Or a lack of proper understanding of the need for improvements, and knowledge and experience of practical measures which can be adopted, constitutes a major obstacle to the reduction of post-harvest losses of fresh produce. However, unlike the factor of price, ignorance can be overcome by a concerted extension and training programme.

In recent years, there has been considerable attention directed to agricultural diversification particularly towards the production of fruits, vegetables and root crops for extra-regional markets. Various workshops and seminars have been held to sensitive extensionists, exporters and producers to the need for improvement in marketing and distribution skills and practices.

C. Focus of Market-Led Extension

Therefore, Marketing extension emphasizes the training of the growers and other marketing functionaries in the following major areas of vegetable marketing:

1. Marketing Functionaries

Madhya Pradesh produce, consume and sell many different crops but they also purchase considerable quantities of produce from the plains in order to meet the requirements of the supplying the needs of the local population during the lean periods when produce would otherwise be scarce. Many people forget all of the vital ancillary services and functions from financial institutions such as banks and credit unions who supply the credit for production as well as distribution activities, transportation, etc. The list of who is involved in marketing include vegetable growers, primary markets, secondary markets, retailers, commission agents, palledars, wholesale-cum-commission agents, contractors, transporters and brokers from outside the region as well as identified by Arora *et al.* (1999).

2. Marketing Channels

The producer of fresh produce in the state is faced with a variety of different options or marketing channels for their crops. Small farmers regularly grow small volumes of many different crops for subsistence purposes and then try to sell any surplus volumes on the domestic market. The consumers in this instance could be other farmers, friends and extended family, or other purchasers in the capital cities and major towns. Seasonal gluts are, therefore, common phenomena in all assembly markets.

In a study by Arora and Saxena (1999) marketing cost of different vegetables was found to be Rs 97.07 per quintal of potato to Rs 112.25 per quintal for cabbage in Kumaon hills. Tripathi and Sharma (1999) found that average marketing cost of Rs 84 per quintal for green pods of peas excluding commission agents' commission in Dehradun market.

In Uttaranchal hills Tripathi and Sharma (1990) found 3 important marketing channels in the marketing of off-season pea. They are (1) producer-consumer (2) producer-commission agent-retailer-consumer (3) producer-local contractor/ forwarding agent-commission agent-retailer.

Arora and Saxena (1999) also found four marketing channels in Kumaon hills. They are (1) producer - contractor - commission agent - retailer - consumer (2) producer-contractor - wholesaler cum commission agent (WCA) - retailer -consumer, (3) producer - commission agent- retailer - consumer (4) producer - WCA - retailer - consumer.

Kumar (1991) suggested that Producer-Primary wholesaler-Secondary wholesaler-Retailer-Consumer channel was more efficient in case of tomato and capsicum whereas in case of beans and peas, Producer-Forwarding agent-Commission agent-Wholesaler-Retailer-Consumer channel was more efficient.

Tripathi and Sharma (1999) revealed that in channel producer-commission agent-retailer-consumer, producers get maximum price of green pods of pea. Study Arora and Saxena (1999) revealed that growers share rises from low priced to high priced vegetables and with increased perishability of vegetables the growers share declines and marketing cost incurred by the farmers is very high in the region.

3. Perishability of Fresh Produce

Most fresh produce is highly perishable and if it is to reach the consumer in the right condition it must be marketed properly bearing in mind the most suitable temperature and humidity for each commodity as well as appropriate packaging and handling methods. Because of the large distances, a large proportion of vegetables is lost and/or damaged in post harvest operations, the extent is sometimes as high as 50 per cent. Non availability of proper infrastructure also contributes to the situation.

Causes of Post-harvest Losses

All fruit, vegetables and root crops are still alive after harvest. They contain from 65 to 95 percent water, depending on the type of produce. For example, watermelons contain about 95 per cent water, while potatoes, yams and other starchy root crops are from 65 to 70 percent water. They also contain food materials, which enable living processes to continue. After harvest the continuation of living processes in the produce uses up both the water and stored food.

- Loss of water from the produce resulting into wilting or shrivel.
- Loss of the food materials of the produce.
- Loss of water and food materials results into loss in weight of the produce.

Factors Affecting the Rate of Breakdown and Loss

The conditions to which produce is exposed after harvest govern its rate of deterioration. These are:

1. The temperature of the produce, which is related to the temperature of the environment, and the heat of respiration of the produce.
2. The extent of damage inflicted during market operations, including physical and physiological damage.
3. The moisture content of the environment.
4. The effect of infection by decay organisms (*e.g.* fungi or bacteria).

4. Handling of the Produce

(i) Before the Harvest

The conditions under which the crop are grown, the cultural practices used, the variety of crop grown and when it was planted, will all determine the quality, readiness for harvest, volume, withstanding the post-harvest status of the crop during marketing and distribution process. One cannot improve the quality of any fresh produce after harvest. Thus the ultimate market quality of the produce is determined by the grower.

a. *Production Planning*

Production planning means a lot more than just selecting the crop and variety to be planted. It means planning: the correct time of planting or sowing; crop spacing; type of inter-crop, if any; irrigation requirements; cultivation practices and frequency; pesticide application and frequency; fertilization type, application and frequency; other cultural practices - pruning/training etc., and the time of harvest.

b. Selection of Crops

Previous experience: The selection of the type of crop will be governed by several inter-related factors. If a grower has no previous experience of a crop he would be foolish to plant more than a few plants on which to experiment.

Soil factors: The grower's soil and fertility will be deciding factors. For example, carrots and sweet potatoes give a higher yield with fewer post-harvest problems if grown on lighter soils that have not received heavy manuring immediately prior to planting whereas bananas require rich fertile loamy soils that have good water absorption properties but which drain freely.

Investment capacity: Vegetable crops occupy land space for relatively short periods, but tree-crops are a more permanent investment and it may be a question of years rather than weeks or months before the first harvest of fruit is obtained.

Seasonal conditions: Seasonal conditions are probably one of the major deciding factors influencing crop selection. Many vegetable crops grow better with fewer pest and disease problems in the dry season provided they have adequate water and overcast skies do not shut out too much sunlight.

c. Varietal Selection

Varietal selection is frequently made on the basis of yield potential, disease resistance, or suitability for local conditions, and often without sufficient consideration of market preferences and post-harvest behaviour. Seed availability of the correct variety is still a very limiting factor. The selection of the variety should follow a clear order of priorities: market preferences, favourable post-harvest characteristics, yield status and agronomic factors, etc.

d. Production Practices for Successful Marketing

Production practices require as much planning as other factors because they often involve capital and labour costs, but more importantly because they influence the production quality of the produce and thus the potential post-harvest life and marketability of the crop.

i. **Irrigation:** Over-irrigation of cucumbers and melons increases fruit size but leads to greater losses due to decay, both before and after harvest, and reduces the intensity of the fruit flavour. Under-irrigation of the same crop improves fruit flavour, but yields are reduced and growth cracks are common. Over-irrigation of leafy vegetables prior to harvest may increase their turgidity but will also make the leaves more brittle and susceptible to handling damage.

ii. **Cultural practices:** Crops growing in competition with weeds may suffer from both water stress and mineral deficiency leading to lower yields and quality. Weeds may also harbour pests and diseases which attack the crop, detract from its appearance and increase the incidence of post-harvest spoilage. Failure to prune out jagged, dead wood in citrus trees makes harvesting difficult, harbours decay organisms and causes physical injury to the fruit, both before and during harvesting.

iii. **Fertilization:** Soil fertility has a direct effect on all aspects of crop growth and development. In some cases, post-harvest disorders can be linked directly to the deficiency of a particular mineral, but often other environmental factors such as water stress are involved. "Spongy tissue" symptoms in mango have been linked to mineral deficiency, and copper and iron deficiencies cause abnormal peel development in citrus fruits. Boron deficiency in the cauliflower affects the yield considerably. In general, fruits and vegetables growing under low fertility conditions are slow to mature, have a greater tendency to develop abnormal shape, do not store well, and ripen irregularly after harvest.

iv. **Chemical treatments:** Pesticides — including insecticides, fungicides, and nematicides, are applied primarily to protect the crop and should therefore improve its quality potential. By removing weed competition, which may impose water stress and mineral stress, herbicides can also have beneficial effects on post-harvest behaviour.

Growth regulators — have a widespread use on fruit crops, causing physiological changes which will improve marketability.

v. **Maturity for Harvest:** A major factor influencing this decision is the price which can be obtained at any given time, which frequently leads to crops being harvested before or after they are at their best in order to get a good price. Two characteristics are normally used in deciding crop maturity, both very dependent on the experience and personal opinion of the grower.

Observation, or looking for characteristic signs of readiness such as:

(a) size and shape (b) colour (c) texture, hardness and softness, etc.

Sampling of characteristic specimens and testing by: a). smell b). taste

Some crops, such as cabbage and yam, are acceptable for consumption over a wide range of development and selection for harvest is often more dependent upon price and the size preferences of the market.

(ii) Post Harvest Treatments

There are some basic pre-treatments which must be performed before storage and/or marketing of any fresh produce.

Cleaning - all stones, soil clods and plant debris must be removed before storage, particularly if the crop is to be stored in bulk.

Grading and selecting - small, damaged, infected and over-mature produce must be removed. Field heat removal - regardless of the type of storage facility employed, it is important to remove the 'field heat' from the produce before bulking up the produce in a store. This field heat removal may be carried out by temporarily stacking the produce in a shaded, cool, ventilated area, or more usually by resorting to refrigeration techniques.

Need of Special Treatments

Amongst vegetables root crops are often 'cured' to prolong their storage life and minimise losses, while onions and potatoes may also be treated with sprout suppressants prior to long-term storage. Fungicides are now in widespread use for decay control in many fruits, vegetables and root crops. Some special treatments are: Ripening and de-greening of fruits, Curing of root crops and onions, Sprout inhibition, Fungicide application, Hot water treatment and Fumigation and vapour-heat.

5. Market Information

Market information includes many aspect of information ranging from production to the final point of sale which improves our ability to market effectively. Some areas of information required on both domestic and distant markets are:

- ✰ size and location of supplies;
- ✰ consumer demand, changes in consumer preference and taste;
- ✰ seasonal variations in supply and demand;
- ✰ historical, current and anticipated prices;
- ✰ selection, grading and packaging requirements;
- ✰ transport availability and charges;
- ✰ phyto-sanitary regulations;

Need of Market Information

Market information and intelligence helps producers, traders and consumers balance supply and demand in the marketing system for fresh produce and thus avoid gluts and deficits in supply and corresponding price fluctuations. Farmers need information about probable supplies and prices in order to make decisions when planning their production, harvest and sale of produce. The knowledge that a farmer can compare one price offered by a trader with another price elsewhere also influences buyers in offering fair prices.

Access to better information enables wholesalers to develop those consumer demands and producer supplies which might otherwise have been neglected. This reduces their business risks and enables them to operate profitably on lower margins. This in turn brings benefits to both producers and consumers. Consumer purchases can also be influenced by market news in that they may choose not to buy expensive produce which is in short supply in favour of more plentiful and cheaper alternatives. Market information must be presented in a simple style which can be understood readily by all those that need to use it.

For the local market most farmers are aware of what crops consumers require, how they expect them to be presented, and the time of year when they are required. This information is not usually consciously acquired - it is simply known from a farmers' general experience of producing and selling. Thus the average farmer knows a great deal about local market conditions, and, if he intends to be successful, must plan his production accordingly. The difficulty which confronts many market information services regarding domestic market information is telling their clients something that is needed and not already known.

Conclusion

At this outset it is much evident that marketing extension seems a new responsibility and challenges for the extension functionaries in the state. For its proper implementation, there is need for proper training of the extension staffs and then intensive training programmes of the vegetable growers along with other market functionaries of the state. The training programmes could be in three phases- first phase for awareness about marketing process, second phase-demonstration of the benefits, third phase-adoption and impact assessment. Therefore, along with the extension scientists, extension workers, planners and NGOs could pay great attention for wide publicity and popularizing the concept.

References

Abbot, J. C. (1970). Marketing fruit and vegetables. FAO Marketing Guide No. 2, 2nd edition. FAO, Rome 181 pp.

Abott, J. C. (1979). Agricultural economics and marketing in the tropics. Intermediate Tropical Agriculture Series. Longman Group Ltd., London 168 pp.

Arora, V P S and Saxena, Raka (1999). Vegetable marketing in hill region of Uttar Pradesh: Status, nature and directions of market reforms. Agricultural Inputs and output Market reforms in India (Conference Proceedings).

Burton, W. G. (1982). Post-harvest physiology of food crops. Longman Group Ltd., London 339 pp.

Dixie, G. (1988). Horticultural marketing. A resource and training manual for extension officers. FAO. 110 pp.

Giuliano, M. A. (1987). Fresh and freign: A complete guide to marketing produce from the Caribbean and Latin America. Walker A. Williams and Company, Inc., Washington DC 76 pp.

Haard, N. F. and Salunke, D. K. (1975). Symposium: Post-harvest biology and handling of fruits and vegetables. AVI Publishing Co. Inc., Westport, Connecticut. 193 pp.

Harris, S. R. (1986). Improvement of post-harvest fresh fruits and vegetables handling. A manual. FAO Regional Office for Asia and the Pacific, and Association of Food Marketing Agencies in Asia and the Pacific. 229 pp.

Srinivas, K. and Bisht, KKS (2003). Status Report on Market trend, consumption pattern, consumer preferences, internal trade, export potential of vegetables produced in hills. Unpublished document, VPKAS, Almora.

Chapter 19

Content Management in Mobile Based Agro-Advisory Services: Status and Methodology

Yagya Dev Mishra and Reeta Mishra

Directorate of Extension Services,
Rajamata Vijayaraje Scindia Krishi Vishwa Vidyalaya, Gwalior, M.P.

Introduction

Despite of incredible growth in service sector during last few decades and noteworthy development in infrastructure and manufacturing sector in our country, the agriculture sector remains an important and promising sector of Indian economy. Self reliant modern agriculture system lay the foundation of food and nutritionally secure nation. Agriculture and allied sectors are still on top in providing employment to the rural and semi urban masses. During last six decades, the efforts in agricultural research, extension and momentous role played by the farmers have resulted in notable increase of food production from 50 million tonnes in 1950-51 to a land mark achievement of 259.32 million tonnes in 2011-12 (MoA, GoI, 2013). The total demand for food grains is projected to touch 280 million tonnes by the year 2020-21. To address this increasing demand, we necessitate a growth rate of nearly two per cent per annum in food grain production (Singh, 2011) and to achieve that the agriculture sector needs to grow at a targeted four per cent per annum.

In present day agriculture, soft resources like knowledge and skills are as important as hard resources like inputs and machinery. In India, we have about 120 million farm holdings and this number is increasing every year. The estimates indicate that 60 per cent of Indian farmers do not have access to any source of information for advanced agricultural technologies resulting in a huge adoption gap. It is estimated that to make available one extension personnel for 800-1000 farm families, at least 1.3 to 1.5 million field level extension personnel will be required, against the present availability of 0.1million extension personnel. Hence, there is an immediate need of vibrant, dynamic and innovative approach to be adopted for agricultural extension in order to achieve targeted growth rate and serve the farmers in a better manner. Further, land and water resources are almost reaching their limits; hence, achieving food security heavily relies on "Knowledge Resource" (Saravanan, 2011). In the current scenario, it is expected that the integration of ICTs in traditional agricultural extension would make available much needed impetus to agricultural sector through delivery of necessary information to the millions of the farmers on regular basis with minimal expenditure. Here the mobile phone comes into the picture with a spark of hope. In today's world, almost everybody owns a mobile phone. This huge reach, if harnessed in agricultural extension, can change the face of agriculture altogether in a developing country like India where we have nothing to lose by using it as a medium to disseminate agricultural information in multimodal form.

Reach of Mobile Phones: Current Status

The fundamental and strategic reforms in telecommunication sector in India after 1990 facilitated a widespread ICT network across the country. The last decade witnessed the rapid stretch of mobile phones in the country. In the beginning of the twenty first century, possessing a mobile phone was a status symbol and a costly choice for an average Indian. But, during last few years, it has created history in fast spread of any goods or technology. In a short span of time, the mobile phone became an essential article for every person of the society from elite class to poor masses. The phenomenal growth of mobile phones surpassed the growth on other popular ICT tools like radio and television. The penetration of mobile phones in rural areas is still less as compare to urban India. Despite of several efforts at Government and private level, there is a huge gap in tele-density of rural and urban localities (Table 19.1).

Table 19.1: Density of Telecom Subscribers (Wireless) in India

Sl.No.	*Area*	*No. of Subscribers (Millions)*	*Tele Density (per cent)*
1.	Rural	414.18	47.78 per cent
2.	Urban	555.71	143.08 per cent
	Total	969.89	80.16 per cent

Source: www.trai.gov.in

Magnitude of Mobile Phone in Agricultural Extension

In the era of information, we like to exchange the information with friends, professionals, clients, etc. for upgrading our knowledge and skills and updating ourselves with the recent developments in every facet of life. Information is thus emerging as more and more powerful tool of governance in our day-to-day endeavour. Experts in the field of information are confident in predicting that our poverty line will no longer be measured in terms of money, but in terms of information access and use. It is merely possible because of the revolution of information communication technology which is responsible for wide spread penetration of computer technology as well as mobile services in to the social fabric. The technology in turn influenced the society, development and social environment. In this age of information revolution, information technologies are being used in almost all walks of life.

Agriculture extension is no longer unscathed with the penetration of ICTs for keeping the farmers updated with research and development in the field of agriculture and allied sciences. The ICTs in general and mobile phones in particular are extensively being used by the public and private extension machinery for disseminating relevant and timely messages to the farming community. Emerging mobile based information and communication networks have redefined the way knowledge is shared, processed, stored and retrieved. Even in rural India, mobile phone is having a profound impact changing the rationale of development activities and strategies. A number of initiatives have proved the potential of it in meeting the information and communication needs of the people for agriculture and rural development.

- ✰ The mobile based messaging enables us the faster sharing of information and innovations and proving as a change agent in changing farmers' lives and agrarian situation by improving access to information and knowledge with negligible time and money involved.
- ✰ It helps in bridging the digital divide situation between haves and have-nots.
- ✰ More than forty different efforts by the Government and private sectors have been made so far in the country for reaching the farmers through mobile phones.
- ✰ The Kisan Mobile Advisories are being issued by various agencies on varied area of interest of the farmers viz; improved agricultural practices, pest and disease control, market and weather updates etc.
- ✰ The Indian Council of Agricultural Research (ICAR) has made it mandatory to issue regular advisories to the farmers and field extension functionaries through its wide spread KVK network and other functional extension units.
- ✰ It helps the farmers in taking day-to-day decisions and developing the risk bearing and entrepreneurial ability.
- ✰ The results of the efforts are significant and encouraging for replications in the future.

Major Public and Private Sector Initiatives

Showing in Table 19.2.

Existing Status of Agro Advisory Services through Mobile Phones

During past few years, the usefulness of Information and Communication Technologies (ICTs) especially Internet and cell phone have increased significantly.

Table 19.2: Agro-Advisory Services/Experiments Initiated through Mobile Phones

Sl.No.	*Name of Project*	*Initiating Agency*	*Year*	*Operational Area*
		Public Sector initiatives		
1.	Farmer Call Centre (Kissan Call Centre)	Ministry of Agriculture, Govt. of India	2004	All India
2.	Mobile Advisory Services by ICAR-KVKs	ICAR, New Delhi	2008	Concerned district of KVK
3.	SMS Broadcast Service	KVK, Babhaleshwar	2006	Ahmednagar, Maharashtra
4.	Mandi on Mobile Digital Mandi	BSNL (U.P.) and Mandi Parishad, U.P. BSNL in collaboration with IIT, Kanpur	2011	Uttar Pradesh
5.	Market Price by SMS	Rubber Board, India	2011	South India
6.	SMS Service to Farmers	Department of Agriculture, Haryana State, India	2008	Haryana and Chandigarh
7.	Dynamic Market Information (DMI), TNAU-C-DAC, Hyderabad	The Tamil Nadu Agricultural University (TNAU) and Centre for Developmentof Advanced Computing (C-DAC), Hyderabad	2008	South India
8.	vKVK (Virtual Krishi Vigyan Kendra)	Govt. Of India (Krishi Vigyan Kendra)	2010	U.P., Uttarakhand and Karnataka
9.	Annapurna Krishi Prasaar Seva:	Govt. of India and ANGRAU, Hyderabad	2012	Andhra Pradesh/Telangana
10.	Mobile based Agro-Advisory System in North-East India (m4agriNEI)	Central Agricultural University (CAU) and Media Lab Asia	2012	Meghalaya
11.	mKisan Portal	Government of India	2013	All India
12.	Intelligent Advisory System for Farmers (IASF)	Department of Information and Technology, Government of India and other collaborating partners	2013	Manipur and Meghalaya
13.	Kissan Kerala	Department of Agriculture, Government of Kerala and Indian Institute of Information Technology and Management – Kerala (IIITM-K)	2009	Kerala
14.	Kisan Help Line	Bihar Agricultural University, Sabour	2012	Bihar
15.	Mobile Multimedia Agriculture Advisory System (MAAS)	ICAR -NAIP	2009	Farmers of India
16.	Mahaagri SMS	Govt of Maharashtra	2009	Maharashtra

Contd...

Table 19.2–Contd...

Sl.No.	Name of Project	Initiating Agency	Year	Operational Area
		Private sector initiatives		
1.	Lifelines India	One World South Asia (OWSA) with the support of British Telecom and CISCO, Tech Mahindra and WIPRO	2006	North and Central India
2.	Fisher Friend	M. S. Swaminathan Research Foundation (MSSRF), Tata Teleservices and Asute System Technology	2007	Tamilnadu
3.	Indian Society of Agribusiness Professionals (ISAP)- Query Redress Services (QRS)	One World South Asia (OWSA) and funded by CISCO and British Telecom		Himachal Pradesh, Madhya Pradesh and Uttar Pradesh
4.	IFFCO Kisan Sanchar Limited (IKSL)	Bharti Airtel Limited and IFFCO	2008	All India
5.	Reuters Market Light (RML)	Reuters Group Plc.	2007	13 states of India
6.	Farmers' Helpline by Chambal Fertilisers and Chemicals Limited (Hello Uttam)	Chambal Fertilisers and Chemicals Limited	2000	Rajasthan, Punjab, U.P. and Haryana
7.	mKRISHI by TATA Consultancy Services	TATA Consultancy Services	2009	Rajasthan, Punjab, U.P., Maharashtra, Gujarat and Andhra Pradesh
8.	Nokia Life Tools	Nokia and Idea cellular	2009	All India
9.	KHETI (Knowledge Help Extension Technology Initiative)	Sironj Crop Producers' Company (SCPCL) and UK Engineering and desktopand Physical computers and Science Research IVRS Council (Private)	2008	M.P.
10.	Behtar Zindagi (Better Life)	Handygo	2012	South and Central India
11.	Awaaz De (Give your voice)	Awaaz De Information Syetem Private Limited	2010	Marginal and Small farmers
12.	Mrittikka	Grameen Intel Social Business Limited and Kutir social Business, Orissa		Eastern India
13.	Fasal	Intuit	2008	Gujarat, Andhra Pradesh and Karnataka

Contd...

Table 19.2–Contd...

Sl.No.	Name of Project	Initiating Agency	Year	Operational Area
14.	KRIBHCO Reliance Kisan Ltd. (KRKL)	Krishak Bharati Cooperative Limited (KRIBHCO) and Reliance Communication Limited	2009	Rural Population
15.	Videokheti	Microsoft Research India and Digital Green		Any farmer and Expert
16.	Mandi Bhav	BSNL, Tata Teleservices and Impetus Technologies	2009	All India
17.	Kisan Sanchar	Srishti Gyan Kendra, ICAR and SAUs	2008	Seven states of India
18.	Warana Unwired	Warana Cooperative, a sugarcane farmers' group and Microsoft Research India	2006	Sugarcane Farmers of Maharashtra
19.	Kissan SMS Portal	Government of India	2013	
20.	mFMS	2012		
21.	Nano Ganesh	Ossian Agro-Automation Pvt. Ltd, India	2008	Mobile based remote control system of water pumps for farmers

These emerging tools are being success in bridging the gap between scientific know how and application of technology at field level by developmental agencies throughout the world. The revolutions in information technologies have a wide and transforming impact on the fields directly linked to our daily needs such as agriculture, healthcare, education, defence and so on. The agricultural extension system is heavily being transformed through the application of internet and mobile phones. The ICT mediated extension systems are acting as key agents for changing agrarian situation and farmers' lives by improving access to information and sharing of knowledge.

As per the statistics of TRAI, the mobile density in rural areas of India is approaching 50 per cent with more than 414 million mobile subscribers in 2015. But the penetration of internet is still not significantly associated with number of mobile owners. The rural people are largely using mobile phone with basic features like voice calls and text messaging. The application of mobile phone in agricultural extension is growing, and has a great potential to be further developed and spread as evident through the success of Kisan Mobile Advisory Services initiated by various Government and private institutions. The Kisan Mobile Advisory services through messages is being used to deliver the needful agricultural information and specially to improve farmers' agricultural technical knowledge with decision making ability, so that, they may enable to increase their production and productivity to fulfil market demands with securing better quality life and income in present competitive agrarian economy. The advisory was sent to targeted farmers covering the broad category of information like, crop production, livestock management, weather forecast, marketing, general awareness and other enterprises etc.

The information delivered through mobile phones is cost effective, timely, specific and personal in nature. Nevertheless, the posing challenges before the scientists, extension professionals and ICT experts engaged in designing and delivering mobile based advisories is for the farmers that how they design, validate, mould, and personalize the contents as per the micro level and diversified needs of the clients. The standard module for the content management in agro-advisories emerged after several trials have been elaborated hereunder.

Content Management Process in Agro-advisories

The quality, timeliness and applicability of content are the key of success for any agro-advisory service.

- ☆ Crispy and timely dissemination of relevant day-to-day information is crucial for farmers
- ☆ Providing control to the communication needs of farmers
- ☆ Converting their Mobiles into Dynamic Power Houses of Knowledge

Essential Components of Agro-advisories

1. Content Planning

Planning is most important step of message designing for communication in any form. The scientists or extension professionals must do content planning for agro-advisories as per the need and level of clients. There are generally two type of messages being spread through mobile advisories.

i) Proactive Message

These are the advisories/messages given to the farmers as per the seasonal requirements in agriculture and allied activities perceived by the service provider or extension agencies. Proactive message includes;

- ✰ Crop Calendar/Package of Practices
- ✰ Cropping Pattern
- ✰ Agro-Ecological Situation
- ✰ Weather Forecast and Market rates

ii) Reactive Message

Reactive messages are normally given in response to some personal queries or announcements in emergency situation. Reactive message includes:

- ✰ Helpline Queries
- ✰ Alerts and Warnings
- ✰ Government Announcements
- ✰ Extreme Events

2. Content Sourcing

For generation of content and material, the public sector mainly dependent on their research units (like universities and ICAR institutes etc.) but private sectors service providers normally employ dedicated teams to generate location specific information and some of them engaging in partnerships with others who can provide better content. Following points should always keep in mind while choosing the source of technology or recommendations for mobile based advisories;

- ✰ Always select latest versions/edition of publications for recommendations and avoid referring literature older than five years.
- ✰ The government policies and preferences in terms of crops/livestock/ enterprises should be given due importance in advisories.
- ✰ For proper management of available sources of content, they should be properly listed, categorised and classified.
- ✰ Get the content source properly approved by the competent authority to avoid legal and administrative hurdles if emerge in later stages.

Few Indicative good sources of content are given hereunder:

- ✰ Refer various publications and websites of Government departments
- ✰ Publications from Agricultural Universities, Research Institutions and Recognised organizations
- ✰ Recommendations from reputed NGOs and established farmers' associations
- ✰ Validated traditional wisdom

3. Content Drafting

Drafting or tailoring of content is one of the most important steps in designing advisories for the farmers. Drafting requires a deep understanding of level of the clients in terms of their age, education, knowledge of language/dialect, agro-ecological situations and many other socio-cultural facts. Without knowing the ground realities of the audience, the delivered content may not give the desired results. Few important remembrances for content drafting are given hereunder.

a. Collect Information from the Right Sources

1. Content Plan/Message Calendars
2. Cropping Pattern (of state/region)
3. Protocols
4. Approved content sources

b. Balance between Scientific Facts and Communication Clarity

1. Explain jargons in simple language
2. Elaborate recommendations, never assume or imply anything
3 Use action verbs and active voice in narration
4. Never recommend banned/restricted chemicals
5. Promote best practices such as INM, IPM etc. whenever possible

c. Clear communication

1. Use interesting narratives, not just dry facts
2. Use locally popular and relevant words to explain scientific terms
3 Messages should raise curiosity in farmers' minds
4. Gender messages should be contextualized keeping in mind local customs and cultures
5. Think of Farmers' First.

d. Proper Understanding of Protocols and Policies

Loyalty to protocols and policies is important. Because, it will provide quality and standard procedure to write a script. Few important and available policies and protocols are listed here below:

1. Gender policy
2. Weather policy
3. Marketing policy
4. Chemical protocol
5. Quality assurance criteria

Chemical Protocol

The chemical protocol should essentially be understood by the content management for effective; result oriented and hazard free communication. The major objectives of chemical protocol are as follows;

- To ensure that chemical pesticides (insecticides, rodenticides, fungicides, and weedicides including plant growth regulators) registered under the Insecticides Act, 1968 by the Registration Committee constituted by the Department of Agriculture and Cooperation, Government of India are used on the crop for the control of insect pests/diseases/weeds; as on the date of recommendation.
- Chemical pesticides recommended are effective against the respective pathogen/insect/weed present in a specific crop.
- Pesticides are prepared, applied and disposed of in a way so as to ensure food safety, health safety and environment safety.

Content for Chemical Protocol

- Number of Applications
- Time of application
- Manner of Application (*i.e.* spraying, dusting, drenching etc)
- Best machinery/tools for application
- Pre-harvest Interval and Maximum Residual Limit
- Safe Handling practices
- Safe disposal of containers

Points to Ponder

Your message is going to reach farmers as an expert advice, hence it should be:

- Impactful, so make it actionable
- Comprehensible, so make it simple
- Interesting, so make it narrative
- Use of actionable, simple and narrative languages

4. Content Quality

The messages delivered to the farmers are audited for its quality based on following parameters:

- *Adequacy*: Is the information actionable?
- *Accuracy*:
  - Is the information scientifically correct
  - Has the information been collected from a legitimate/approved source
  - Adherence to Protocols (Use of Chemicals, Gender issues, Weather information, Market information)
- *Relevancy*:
  - Does the information adhere to the Crop Calendar/cropping system
  - Does the information adhere to the Message Mix/message calendar
  - Adherence to Content Plan

☆ ***Simple and clear language:***
  - Information does not contain jargons
  - Confident and clear voice (for voice message)

5. Content Dissemination

Like other components of content management, proper dissemination of advisories is equally important for timely action and desired results. Following aspects should be considered thoughtfully while disseminate the content to the farmers.

☆ Text or Voice message: Based on client

☆ Size of message: Voice message duration and length of text message

☆ Frequency: Number per day

☆ Time of delivery: As per the actual need of information at that point of time

☆ Relevant Information: As per requirements

☆ Authentic/Reliable Source of Content: Credibility for long term relationship

☆ Measurability of response and monitoring: Feedback through regular monitoring for improvement

6. Content Management/Documentation

Documentation is essentially required for future reference. Documentation can be done in electronic and printed forms.

Essentials of Documentation

☆ Proper classification and categorisation of the messages

☆ Understanding knowledge management system

☆ Inter-portability of Knowledge
  - *E.g.* usages of English synonyms/meaning for local terms
  - Universal measurement system
  - Reference

☆ Complete documentation of the messages released/broadcasted

☆ Coding for matching text and sound files in the system

☆ ***Key words:*** *The key words are used as Tags hence to be chosen carefully. Ideally there should not more than two key words per message. Following should be used as keywords;*

a. Crop Production:

i. Name/s of fertilizers

ii. Name/s of plant growth promoters (if mentioned)

iii. Name/s of special agri-implements

iv. Critical agro-climatic conditions/effects (frosting, scorching etc.)

v. Typical practices (SRI, zero tillage, IPM etc.)

vi. Typical varietal features (GMO, Hybrid etc.)

b. Plant Protection:

i. *Chemical name/s of pest control agents (Insecticides, fungicides, herbicides etc.)*

ii. *Common/Commercial Names of pest control agents (must be at least two form two different manufacturers in order to avoid implicit favours)*

Proper Categorisation of the Messages

Classify messages significantly for maintaining strong content repository.

Why

- ☆ Helps to maintain proper documentation
- ☆ Helps in audit
- ☆ Helps in referring if required later
- ☆ Content disseminators from other places can also reuse it

How

- ☆ Understand significance of each field.
- ☆ Select correct fields from drop down menu
- ☆ As far as possible do not leave any field blank.

Understanding Content Categorisation

Reference Value

- ☆ Messages/queries which add to the knowledge domain of the listener and improve his/her decision making ability, are included here.
- ☆ All general crop production practices like land preparation, sowing, weed management, water management etc in crops and feeding, housing health care etc. in animal husbandry are categorized under this group.

General Announcement/Alert

Messages and Helpline queries on most non – agriculture related subjects, events, government schemes should fall in this category.

Routine

Weather, Mandi rates messages which are given regularly and queries on these, which have only one time value should form part of this category.

Checklist for Content Documentation

- ☆ Use simple terms
- ☆ Mention English meanings for local terms used in messages
- ☆ Use actionable words
- ☆ Use same measurement units
- ☆ Mention complete transcription
- ☆ Classify/Categorise the messages properly

Common Mistakes while Documentation

- ☆ Incomplete documentation
- ☆ Sentences are not properly framed
- ☆ Use of unknown local terms
- ☆ Improper categorisation of subject matter

Principles of good Knowledge Management Practices

- ☆ **Situational:** changes with target audience, users' need and time
- ☆ **Evolving**: discovering new information and creating knowledge assets
- ☆ **Interoperable**: can be shared between multiple users and learned by different profiles of users
- ☆ **Quality assured:** embedded quality assurance process to ensure knowledge created authentic, accurate and reusable.

7. Feedback

Feedback from farmers/client is equally important for effective content management. Feedback helps in knowing the status of adoption, preference of subject areas, difficulties in understanding and appropriate time of delivery which ultimately helps in improving the overall status of the services.

Key Considerations in Mobile Based Agro-advisory Services

Mobile phone is a fastest reaching tool of communication irrespective of area, state, age, education and other socio-economic factors. It is available in diverse choices from basic features (talk and text messaging) to smart applications (multimedia and internet features). It has emerged as an extensively available option for providing multiple offers and opportunities for day-to-day communication and connectivity with farming community. The mobile based advisories need to consider following points;

- ☆ In mobile advisory services, equal partnership of content generator, disseminator and user is an important key of success.
- ☆ Farmers normally give better response to the content designed with local flavour, explicit and personalize counselling.
- ☆ In addition to text messages, voice, image and multimedia based services need to be initiated to sustain momentum in the mobile advisory services.
- ☆ To ensure sustainability of services, innovativeness and creativity in content delivery is having ample importance.
- ☆ Time of delivery is another important consideration in consideration for impactful advisory.
- ☆ Credibility and quality of the message sent to the farmer, irrespective of the format, determines success of the service to a great extent.

References

Agrawal, S., Singh S.R.K. and Rajan, P. (2014). Correlation of technical knowledge of Kisan Mobile Sandesh beneficiaries in Jabalpur district: A case of mobile-based ICT application, Agriculture Update, Vol. 9(2). pp 237-242.

Ahuja, Vivek (2011). Cyber extension: A convergence of ICT and agricultural development, Global Media Journal- Indian Edition/ISSN 2249-5835, Vol. 2 (2).

Bhaskar, G. (2013). Mobile SMS Application in Agricultural Information Dissemination: A Case on KVK, Bhabaleshwar SMS Alerts, Reading Material,Training Program on Applicationof ICTs in Modified Extension Reforms, National Institute of Agricultural Extension Management (MANAGE), Hyderabad.

Dey, B.L., Prendergast, R. and Newman, D.: How can ICTs be used and appropriated to address agricultural information needs of Bangladeshi farmers?, Available at web link: http://www.globdev.org/files/28-Paper-Dey-How per cent 20Can per cent 20ICTs per cent 20be per cent 20used-Revised.pdf

Glendenning, C.J., Babu, S., and Okyere, K.A. (2010). Review of agricultural extension in India- Are farmers' information needs being met?, IFPRI Discussion paper 01048, Eastern and Southern Africa Regional Office.

Kumar, S., Singh, S.R.K. and Sharma, R. C. (2014). Impact of Kisan Mobile Advisory Service on Transfer of Agricultural Technologies, International Journal of Extension Education, V ol. 10, pp. 70-72.

Lio, M. and Liu, M. (2006). "ICT and agricultural productivity: evidence from cross-country data", Agricultural Economics, Vol. 34, pp. 221-228.

Mittal, Surabhi, Gandhi, Sanjay and Tripathi, Gaurav, (2010). Socio-Economic Impact of Mobile Phones on Indian Agriculture. ICRIER Working Paper No. 246, Indian Council for Research on International Economic Relations (ICRIER), New Delhi.

Patel, M. R., Patel, M. V. and Patel, R. A. (2015). Assessment of Kisan Mobile Advisory (KMA) Service for Dissemination of Agriculture Information in Mehsana District; Gujarat, International Journal on Recent and Innovation Trends in Computing and Communication, Vol. 3(7), pp. 4599-4602.

Parganiha, O., Shrivastava, S. K., Chaubey, A. K. Nag, J. L. (2012). Impact of Kisan Mobile Advisory (KMA) On Agricultural Technology Dissemination, Indian Research Journal of Extension Education, Special Issue (Vol. - II). pp 175-178.

Qaisar, T. M., Khan, M.M.A., and Alam, S. (2011). Innovative Agricultural Information Services by ICT Projects in India, International Journal of Trade, Economics and Finance Vol. 2 (4), pp. 276-279.

Ramaraju, G. V. Anurag, T. S, Singh, H.K., Kumar, S. (2011) ICT in Agriculture: Gaps and Way Forward, Volume 21, No. 2, July 2011.

Saravanan, R. (2010). ICTs for Agricultural Extension: Global Experiments, Innovations and Experiences, New India Publishing Agency, New Delhi.

Saravanan, R. (2010). ICTs for Agricultural Extension in India: Policy Implications for Developing Countries, School of Social Sciences, College of Post Graduate Studies, Central Agricultural University (CAU) India.

Saravanan, R. (2011). ICTs for Agricultural Extension in India: Innovations, Lessons and Way forward. Presented as an Invited panellist for the APAARI Session on Openness in Agricultural Information and Knowledge Sharing during the INSEE International Mobile Phone Applications for Agricultural Extension in India 73 Conference -2011 on "Innovative Approaches for Agricultural Knowledge Management: Global Extension Experiences" organised by the International Society of Extension Education, New Delhi, India.

Saravanan, R., and Suchiradipta Bhattacharjee, (2014). Mobile Phone Applications for Agricultural Extension in India, In: Saravanan, R (Ed.). Mobile Phones for Agricultural Extension: Worldwide mAgri Innovations and Promise for Future, New India Publishing Agency, New Delhi. pp. 1-75

Sharma, Gyanendra (2005). ICT based transfer of technology programme- A review of development initiatives, ICT applications in transfer of agricultural technologies: A reference manual of ICAR sponsored summer school, Department of Ag. Communication, College of Agriculture, GBPUA and T, Pantnagar.

Singh, B.B. (2005). ICT in transfer of agricultural technology, ICT applications in transfer of agricultural technologies: A reference manual of ICAR sponsored summer school, Department of Ag. Communication, College of Agriculture, GBPUA and T, Pantnagar.

Singh, Geeta, Dixit, H. and Pathak, Renu (2013). Kisan Mobile Sandesh Reaches to Unreached, TECHNOFAME- A Journal of Multidisciplinary Advance Research, Vol. 2(1), pp. 54-60.

http://www.iksl.in/

www.**rvskvv**.net

Chapter 20

Migration Pattern in Command Areas of Chhattisgarh: A Farmers Perspective Analysis

M.L. Sharma, A.L. Rathore, and A.K. Gupta

Department of Agricultural Extension,
IGAU, Raipur, C.G.

Death, birth and migration are the key features for maintaining equilibrium in social system. Temporary or seasonal shifting of landless, small and marginal farmers for earnings during off crop season is considered as migration for this study. Migration is not harmful in terms of migrants because if migrants earn money without losing health, social reputation, poverty and education of children. In, general, the migration become a sign of poverty in most part of eastern India, where majority of farmers are tied with mono cropping of rice. This study was carried out during the year 2008-2009 in selected 12 districts of the Chhattisgarh state under Asian Development Bank (ADB) funded project. Out of this 684 farm families only 193 migrated farm families from existing 51 Water Users Associations (WUAs) were considered as respondents for studying "Migration pattern among the farmers of Chhattisgarh: Insights from household Survey in Command area. Data were collected through personal interview method with the help of structured interview schedule and Focus Group Discussion (FGD) with active and knowledgeable farmers of the WUA. While, collection of secondary data was done with the help of the local leaders of the village and village level Government official (Patwari, RAEO, Amine, Panchayat members, Executive members of WUA etc. The study indicate that out of 684 farm families (71.80 per cent) only, 193 farm families (28.20 per cent) had done migration. Average age of the respondents was 45 year and among them 85 per cent were head of the family. Majority (50.4 per cent) of them had joint type of family in which 67.2 per cent of male and 65.2 per cent of female were working members of the family. The average size of family was 7.4 persons. Majority of the respondents (55.1 per cent) were belongs to Other backward caste (OBC) and majority of them (92.33 per cent) were obtaining the information related to migration from friends. The findings describes that majority of the migrants (52.85 per cent) were male only, majority of them (75.13 per cent) were migrate during December to February and 58.03 per cent had migrated for 30 to 60 days. 74.09 per cent respondents were migrated every year, 48.70 per cent were migrated due to stress and most of the migrants (69.95 per cent) were migrated to another village of the district. Majority of the respondents (87.05 per cent) were saying that they were migrated in the search of Better wages and most of them (56.99 per cent) were engaged in agriculture and in some other works like construction/repair (16.06 per cent), industries/factory and paid labour in shop and other (8.81 per cent). With regard to effect of migration on their socio-economic condition the findings reveals that effect on economic status, effect on social status and effect on family opinion had positive percentage effect and the figure was +54.40, +71.76 and +16.06 per cent respectively. While, effect on child education, effect on working efficiency of female, effect on agriculture work and effect on animal husbandry had negative percentage effect and the figure was -17.36, -21.50, -19.17 and -28.50 per cent respectively. During male migration the percentage change in time required for doing different works/activities like domestic work, child care, social participation, animal husbandry, agriculture/other work and other activities was 26.67, 40.00, 00.00, 50.00, 16.67 and -75.00 per cent respectively. Similarly, during male with female migration the percentage change in time required for doing different works/activities like domestic work, child care, social participation, animal husbandry, agriculture/other work and other activities was -35.71, -33.33, -50.00,-100.00, 60.00 and 11.11 per cent respectively.

Introduction

Human migration is the movement by people from one place to another with the intentions of settling temporarily or permanently in the new location. The movement is often over long distances and from one country to another, but internal migration is also possible; indeed, this is the dominant form globally. Migration may be individuals, family units or in large groups (Anonymous, 2014a). A "working migrant" has been defined by the United Nations Convention on the Rights of Migrants as a "person who is to be engaged, is engaged or has been engaged in a remunerated activity in a State of which he or she is not a national".(Anonymous, 2013)

Nomadic movements are normally not regarded as migrations as there is no intention to settle in the new place and because the movement is generally seasonal. Only a few nomadic peoples have retained this form of lifestyle in modern times. Also, the temporary movement of people for the purpose of travel, tourism, pilgrimages, or the commute is not regarded as migration, in the absence of an intention to live and settle in the visited places. (Anonymous, 2016)

Migration occurs because individuals search for food, sex and security outside their usual habitation. In towns and cities there are a creation of the human struggle to obtain food, sex and security. To produce food, security and reproduction, human beings must, out of necessity, move out of their usual habitation and enter into indispensable social relationships that are cooperative or antagonistic. Human beings also develop the tools and equipment to enable them to interact with nature to produce the desired food and security. The improved relationship (cooperative relationships) among human beings and improved technology further conditioned by the push and pull factors all interact together to cause or bring about migration and higher concentration of individuals into towns and cities.

The higher the technology of production of food and security and the higher the cooperative relationship among human beings in the production of food and security and in the reproduction of the human species, the higher would be the push and pull factors in the migration and concentration of human beings in towns and cities. Countryside, towns and cities do not just exist but they do so to meet the human basic needs of food, security and the reproduction of the human species. Therefore, migration occurs because individuals search for food, sex and security outside their usual habitation. Social services in the towns and cities are provided to meet these basic needs for human survival and pleasure. (Anonymous, 2016)

Factors Responsible for Migration

Lee divide factors causing migrations into two groups of factors: push and pull factors. Push factors are things that are unfavorable about the area that one lives in, and pull factors are things that attract one to another area (Lee, 1966).

Push Factors

It includes the factors like not enough job, few opportunities, inadequate condition, desertification, famine or drought, political fear or persecution, poor medical care, loss of wealth, natural disasters, war etc.

Pull Factors

Some pull factors that is responsible for migration are job opportunities, better living conditions, the feeling of having more political and/or religious freedom, enjoyment, education, better medical care, security, family links etc.

In 2013, the number of international migrants born in the South who lived in the North, or "South-North migration", almost equaled the number of migrants born in the South who resided in the South, or "South-South migration". The proportion of female migrants ranged from 52 per cent in the global North to 43 per cent in the global South in 2013. Six out of every ten international migrants under the age of 20 resided in developing regions. Conversely, about seven out of every ten migrants aged 60 and above lived in the developed regions. During the period 2000-10, the global migrant stock grew twice as fast than during the previous decade. During the 1990s, the global migrant stock grew at an average of about 2 million migrants per year. During the decade 2000-10, the growth in the migrant stock accelerated to about 4.6 million migrants annually. However, since 2010, the increase in the migrant stock has slowed down. In the aftermath of the global economic crisis, the annual increase in the global migrant stock fell to about 3.6 million since 2010. (United Nations, 2013a)

The World Bank's Migration and Remittances Factbook of 2011 lists the following estimates for the year 2010: total number of immigrants: 215.8 million or 3.2 per cent of world population. In 2013, the percentage of international migrants worldwide increased by 33 per cent with 59 per cent of migrants targeting developed regions. Almost half of these migrants are women, which is one of the most significant migrant-pattern changes in the last half century. Women migrate alone or with their family members and community. Even though female migration is largely viewed as associational rather than independent migration, emerging studies argue suggest complex and manifold reasons for this. (*UNFPA, 2013*)

The top ten destination countries are USA, Russian Federation, Germany, Saudi Arabia, Canada, UK, Spain, France, Australia and India. Similarly, the top ten countries of origin are Mexico, India, Russian Federation, China, Ukraine, Bangladesh, Pakistan, UK, Philippines and Turkey. (Anonymous, 2016)

Table 20.1: International Migrant Stock (millions)

	1990	*2000*	*2010*	*2013*
World	154.2	174.5	220.7	231.5
Developed regions	82.3	103.4	129.7	135.6
Developing regions	71.9	71.1	91	95.9
Africa	15.6	15.6	17.1	18.6
Asia	49.9	50.4	67.8	70.8
Europe	49	56.2	69.2	72.4
Latin America and the Caribbean	7.1	6.5	8.1	8.5
Northern America	27.8	40.4	51.2	53.1
Oceania	4.7	5.4	7.3	7.9

Source: United Nations (2013), Trends in International Migrant Stock: The 2013 Revision.Data of international migration stock is given below in the Table 20.1. Table depicts that international migration stock was increase by 77.3 million from 1990 to 2013. In developed regions from the year 1990 to 2013, international migration stock was changed from 82.3 million to 135.6 million. While for developing regions the data varies from 71.9 million to 95.9 million for the year 1990 to 2013. In the year 2013 the maximum international migration stock was for Europe (72.4 million) followed by Asia (70.8 million) and Northern America (53.1 million). (United Nations, 2013b)

International migration in India can broadly be classified into two patterns such as migration of Indians to other countries and migration of the nationals of other countries, voluntary or non-voluntary, to India. But despite to these, most frequently observed types of migration in India are given below:

1. Emigration

Migration from India to various parts of the world was very old and in modern times five patterns of Indian emigration were identified such as i) Unskilled labour emigration, ii) Kangani*/Maistry* or Contract labour emigration, iii) Free or Passage21 emigration, iv) Technologically Experts emigration v) Labour migration to West Asia (Jain, 1989). The first three types of emigration were colonial phenomenon but the last two were the result of inherent contradiction of post colonial socio-economic development in India. However, the migratory patterns in India can be classified into four heads.

2. Immigration

Migration of people from different countries, more significantly from neighbouring countries, to India was associated with the democratic polity, liberal society, religious freedom and developed economy coupled with more job avenues. Since 1947 the number of legal as well as illegal immigrants in India was been increasing over the decades.

3. Refugee Migration

There had also been a significant trend of an involuntary or forced immigration to India in the form of refugees during the post independence period. By the end of 2006, over 300,000 nationals of Bangladesh, Tibet and Sri Lanka were recognized as refugees by the Indian government and the United Nations High Commission for Refugees (UNHCR) and were given shelter in India.

4. Internal Migration in India

Internal migration means movement of people from one part to other within the country and it is very predominant in India. Internal migrants in India constitute a large population: 309 million internal migrants or 30 per cent of the population (Census of India 2001), and by more recent estimates 326 million or 28.5 per cent of the population. Out of the total internal migrants, 70.7 per cent are women (Census of India 2001). Marriage is given as the prominent reason for female migration in both the rural and urban areas 91 per cent of rural female migrants and 61 per cent of the urban female migrants. Migration for employment-related reasons is given as the prominent reason for male migration in both rural and urban areas 29 per cent rural male migrants and 56 per cent of urban male migrants (NSSO 2007–2008).

Lead source states are Uttar Pradesh, Bihar, Rajasthan, Madhya Pradesh, Andhra Pradesh Chhattisgarh, Jharkhand, Orissa, Uttarakhand and Tamil Nadu. Key destination states are Delhi, Maharashtra, Gujarat, Haryana, Punjab and Karnataka.Migrants are mostly employed in the following subsectors: construction, domestic work, textile, brick-kilns, transportation, mines, qua rries and agriculture (Deshingkar and Akter 2009).

Chhattisgarh was first recognized nationally as a supplier of cheap labour towards the end of the 19th century when, immediately after the completion of the

---

**Under the Kangani system Indians were migrated to Malaysia and Sri Lanka as contract labourers to work on sugar and rubber plantations. A big number of Indians were also migrated to work on sugar and rubber plantations in Myanmar under the system of Maistry. Such labourers were mainly from Tamil speaking areas in the Madras Presidency.*

Eastern segment of the Howrah-Bombay rail link, thousands of labour from the famine struck plains were carted to the tea gardens of Assam (Anonymous, 2015a). Earlier human migration from present Chhattisgarh had reportedly occurred in the late 19th Century. Much later, drought years during the 1960s had triggered wide-spread migration. However, in the recent years migration has become more of a norm than exception, accentuated by misguided development policies that force the poor to migrate in search of improved livelihoods. (Anonymous, 2014b)

According to government statistics, 95,324 people have left their homes in last three years. A substantial part of the migration 29,190 took place from Janjgir-Champa district. Inquiries revealed that people living below the poverty line are not getting jobs under MGNREGA scheme. Apart from Janjgir-Champa, there are also reports of migration from Bemetara, Balod, Kawardha, Gariaband, Durg and Dhamtari districts, among others. The migrants have moved to Uttar Pradesh, Bihar, Jammu and Kashmir and Madhya Pradesh where they are engaged in construction of houses, roads and canals or they work in brick kilns. (Anonymous, 2015b)

The state government has informed the assembly that nearly one lakh people have left their homes and migrated in search of job in the state in the last three years, conservative non-official estimates put the figure at more than 20 lakhs. The period of lull coincides with the period the previous UPA government pushed its rural safety net programme, MGNREGA, with much enthusiasm; the change has come with the Raman Singh government going slow on it in the state after the Centre reduced allocation to the state government. (Anonymous, 2015a)

Methodology

This study was carried out during the year 2008-2009 in selected 12 districts of the Chhattisgarh state under Asian Development Bank (ADB) funded project. For this study all the existing 51 Water Users Associations (WUAs) were considered for selection of farm families. Farm families were selected on the basis of command area and two farmers were selected from each 100 ha area of WUAs or minimum of 12 farmers from each area of WUA. In this way a total of 684 farm families from all the 51 WUAs were selected. Out of this 684 farm families only 193 migrated farm families were considered as respondents for studying "Migration pattern among the farmers of Chhattisgarh: Insights from household Survey in Command area.

In this study two types of data *i.e.* primary and secondary were collected. Primary data of was collected through personal interview method with the help of structured interview schedule and Focus Group Discussion (FGD) with active and knowledgeable farmers of the WUA. While, collection of secondary data was done with the help of the local leaders of the village and village level Government official (Patwari, RAEO, Amine, Panchayat members, Executive members of WUA etc.

Results and Discussion

Migration Pattern among Farm Families

Migration pattern among farm families is presented in Table 20.2. The findings portrays that majority of the farm families (71.80 per cent) were belongs to non migrated category while, 28.20 per cent were belongs to migrated category. This 28.20 per cent farm families were considered as the respondent for this study. The probable reason behind non migration is that in the state there is ample scope for employment through MGNREGA and assured food security through public distribution system (PDS). While, probable reason behind migration might be because of the facts that migrants are habituated to migrate or their high level of aspiration force them to migrate or they are unable to get employment in the local are because of their limited and scare resources.

Table 20.2: Migration Pattern among Farm Families (n= 684)

Sl.No.	*Particular*	*Frequency*	*Percentage*
1.	Migrated	193	28.20
2.	Non migrated	491	71.80

Socio-Personal Information of Respondents in Chhattisgarh

Table 20.3 reveals about the socio-personal information of migrated respondents. The result shows that Average age of the respondents was 45 year. Majority of the respondents (85 per cent) were head of the family while, 15 per cent were relatives of head of family.

Table 20.3: Socio-Personal Information of Respondents in Chhattisgarh (n= 193)

Particulars	
☆ **Average age of the respondents (years)**	45
☆ **Relation of the respondent with head of family**	
● Head of family	85
● Relatives	15
☆ **Type of family (per cent)**	
● Nuclear	49.6
● Joint	50.4
☆ **Total size of family**	
● **Male**	
☆ **Non-working (0-15 years)**	32.8
☆ **Working (Above 15 years)**	67.2
● **Female**	
☆ **Non-working (0-15 years)**	34.8
☆ **Working (Above 15 years)**	65.2
☆ **Average size of family (per cent)**	7.3
☆ **Caste category of household (per cent)**	
● Schedule caste	11.3
● Schedule tribe	19.2
● Other backward caste	55.1
● General	14.5
☆ **BPL categories of households (per cent)**	
● BPL category	53.8
● Non BPL category	46.2

Majority (50.4 per cent) of them had joint type of family and 49.6 percent were belongs to nuclear family. In respect of total size of family 67.2 per cent of male and 65.2 per cent of female were working members of the family. While, 32.8 per cent of male and 34.8 per cent of female were non-working members of the family.

The average size of family was 7.3 persons. Majority of the respondents (55.1 per cent) were belongs to Other backward caste (OBC) followed by schedule tribe (19.2 per cent), general caste (14.5 per cent) and schedule caste (11.3 per cent). 53.8 per cent of them belongs to BPL category while, 46.2 per cent were belongs to non BPL category.

Parganiha (2002) found that 70 per cent of respondents were belongs to joint family and 30 per cent were belongs to nuclear family.

Sources of Information about Migration

Sources of information about migration are presented in the Table 20.4. The results says that majority of the migrated respondents (92.23 per cent) were obtaining the information related to migration from friends followed by neighbor (90.16 per cent), relatives (84.97 per cent) and local middleman (40.41 per cent).While, 39.90 per cent were obtaining the information from society head, 17.10 per cent from news paper, 15.03 per cent from broker, 12.95 per cent from radio and only 9.84 per cent from village head.The probable result might be because of the fact that they use such sources of information which is easily accessible and utilize by them.

Table 20.4: Sources of Information about Migration (n=193)

Sources of Information	*Frequency**	*Percentage*
Village head	19	9.84
Society head	77	39.90
Local middleman	78	40.41
Friends	178	92.23
Neighbors	174	90.16
Relatives	164	84.97
Broker	29	15.03
News paper	33	17.10
Radio	25	12.95

* Data are based on multiple responses.

Parganiha (2002) found almost similar findings which revealed that most of the respondents were getting the information from the sources like friends, neighbors and relatives.

Migration Pattern Among the Farm Families in Study Area

Table 20.5 represents the data about migration pattern among the farm families in study area. The findings describes that majority of the migrants (52.85 per cent) were male followed by female (20.21 per cent), both male and female (17.10 per cent) and whole family (10.88 per cent). The migration of male members only might be because of that male has the social responsibility of rearing of his family members and due to unemployment in the lean season they are bounded to go outside from village in the search of employment. While, the female members migrated only when they are widow and they have the responsibility of their family rearing lies on their shoulder.

Table 20.5: Migration Pattern Among the Farm Families in Study Area (n=193)

Particular	Frequency*	Percentage
☆ **Migrants**		
● Male only	102	52.85
● Female only	39	20.21
● Both male and female	33	17.10
● Whole family	21	10.88
☆ **Period of migration**		
● October to November	29	15.03
● December to February	145	75.13
● March to April	17	8.81
☆ **Duration (days) of the migration**		
● 30-60	112	58.03
● 60-150	66	34.20
● Above 150	102	52.85

* Data are based on multiple responses.

The majority of the migrated respondents (75.13 per cent) were migrate during December to February followed by October to November (15.03 per cent) and March to April (8.81 per cent). The probable result might be because of the fact that majority of the migrants have no employment in their local vicinity at this lean agricultural period.

The findings of Table 5 also reveal that majority of the migrants (58.03 per cent) had migrated for 30 to 60 days followed by above 150 days (52.85 per cent) and only 34.20 per cent migrants were migrated for 60 to 150 days.

Choubey (2000) found that majority of the migrants (47.83 per cent) were both male and female and most of them migrated for the period between 1.1 to 3 months.

Distribution of Respondents According to their Migration Pattern

Distributions of respondents according to their migration pattern are shown in Table 20.6. It is found that most of the migrants (74.09 per cent) were migrated every year followed by alternate years (15.03 per cent), after two years (7.25 per cent). While, only 3.63 per cent were migrated after more than two years. It might be due to unavailability of employment every year during lean agricultural period.

Table 20.6 also ravels about nature of migration. In this context result depicts that majority of the migrated respondents (48.70 per cent) were migrated due to stress while, 33.68 per cent were due to agreement and 17.62 per cent were follows routine migration. The probable reason may be unavailability of money and food which leads the stress. Sometimes stress like drought increase the significantly rate of migration particularly among the small and marginal farmers. (Sharma, *et al.*, 2002)

Table 20.6: Distribution of Respondents According to their Migration Pattern (n=193)

Particular	Frequency	Percentage
☆ **Frequency of migration**		
● Every year	143	74.09
● Alternate years	29	15.03
● After two years	14	7.25
● After more than two years	7	3.63
☆ **Nature of migration**		
● Due to agreement	65	33.68
● Due to stress	94	48.70
● Routine migration	34	17.62
☆ **Place of migration**		
● Another village of the district	135	69.95
● Another district of the state	27	13.99
● Village/Town/District of another state	31	16.06

With regard to place of migration most of the migrants (69.95 per cent) were migrated to another village of the district, followed by village/town/district of another state (16.06 per cent) and another district of the state (13.99 per cent). Most of the migrants were migrated to another village of the district which indicates that in the study area domestic migration is dominant and occurred more frequently. It can be say that it might be happen because of migrants are preferring to migrates those place which is near to their native place. Parganiha (2002) reported that majority of the respondents were migrated outside the block, stress forces to most of them to migrate and majority of them were migrated every year.

Reasons of Migration and Type of Work Carried Out by the Migrated Respondents in the Study Area

Reasons of migration in the study area are presented in the Table 20.7. It is clearly visible from Table 20.6 that majority of the respondents (87.05 per cent) were saying that they were migrated in the search of Better wages, followed by other reasons like small land holdings, big family, natural calamities etc. (16.58 per cent), lesser employment opportunities in local level (4.15 per cent). While, 2.07 per cent had given the reason of both lesser employment opportunities in local level and setting up a business and 1.04 per cent had told both in search of job and getting additional income.

It can be concluded from this finding that there is unavailability of better job in the local area of migrants. According to Anonymous, 2016 migration occurs because individuals search for food, sex and security outside their usual habitation. Social services in the towns and cities are provided to meet these basic needs for human survival and pleasure.

Table 20.7 also depicts about type of work carried out by the migrated respondents in the study area. the findings portrays that most of the respondents (56.99 per cent) were engaged in agriculture work followed by construction/repair (16.06 per cent), industries/factory and paid labour in shop and other (8.81 per cent), earth working (5.18 per cent) and bricks making (3.11 per cent). While 1.04 per cent were engaged in running of rick sow or auto. It might be because of the reason that they are more skilled in doing work related to agriculture than the other works that requires specific skill to perform it.

Choubey (2000) observed that majority of respondents (39.13 per cent) were engaged in building construction followed by brick making (30.43 per cent), agricultural labour (13.04 per cent) and both bamboo cutting and other work (8.70 per cent).

Table 20.7: Reasons of Migration and Type of Work Carried Out by the Migrated Respondents in the Study Area (n=193)

Particular	*Frequency**	*Percentage*
☆ **Reasons of migration**		
● Better wages	168	87.05
● Lesser employment opportunities in local level	8	4.15
● No employment opportunities in local level	4	2.07
● Setting up a business	4	2.07
● In search of job	2	1.04
● Getting additional income	2	1.04
● Other reasons (Small land holdings, big family, natural calamities etc.)	32	16.58
☆ **Type of work carried out**		
● Agriculture work	110	56.99
● Bricks making	6	3.11
● Construction/Repair	31	16.06
● Earth working	10	5.18
● Industries/Factory	17	8.81
● Rick sow/Auto	2	1.04
● Paid labour in shop and other	17	8.81

* Data based on multiple responses.

Perception of Respondents Regarding the Effect of Migration on their Socio-economic Condition

Table 20.8 portrays about perception of respondents regarding the effect of migration on their socio-economic condition. The findings show that out of 193 migrants 118 were found that their economic status was improved followed by 109 migrants said that their social status was too much improved, 126 said their child education status was decreased and 134 were says that working efficiency of female was decreased.

Table 20.8: Perception of Respondents Regarding the Effect of Migration on their Socio-economic Condition (n=193)

Statement	*Effect on Farmers Status*				*Per cent Effect*
	Too much Improve	*Improve*	*No Change*	*Decreased*	
Effect on economic status	50	118	17	8	+54.40
Effect on social status	109	59	25	0	+71.76
Effect on child education	17	25	25	126	–17.36
Effect on working efficiency of female	17	17	25	134	–21.50
Effect on agriculture work	17	17	34	125	–19.17
Effect on animal husbandry	8	17	25	143	–28.50
Effect on family opinion	34	68	17	74	+16.06

Similarly, 125 migrants were reveals that their involvement in agriculture work was decreased, 143 were said their involvement on animal husbandry was decreased while, 74 were told that their family opinion was decreased.

In the context of percentage effect the result indicated that effect on economic status, effect on social status and effect on family opinion had positive percentage effect. the data of this positive percentage effect was 54.40, 71.76 and 16.06 per cent respectively. While, effect on child education, effect on working efficiency of female, effect on agriculture work and effect on animal husbandry had negative percentage effect. the data for such negative percentage effect was -17.36, -21.50, -19.17 and -28.50 per cent respectively.

The probable reason behind this result is that due to migration the migrants are getting the opportunity of employment hence there economic and social status is improved. But due to migration the involvement of migrants for agriculture work and animal husbandry is decreased and in the case of seasonal migration the education of their child (less than 8 years) is also disturbed. Migration decreased the working efficiency of female because they are engaged in both wage earning and caring of family members.

Perception of Respondents Regarding the Effect of Migration on Female Work Load

Perception of respondents regarding the effect of migration on female work load is given in Table 20.9. The findings indicate that total time required (hr/day) for domestic work was 7 hrs. Similarly, for child care 2 hrs, social participation and animal husbandry 1 hr, agriculture or other work 6 hrs and for other activities it was 7 hrs.

Table 20.9: Perception of Respondents Regarding the Effect of Migration on Female Work Load (n=193)

Work during Migration	*Total Time Required for Work (hr/day)*	*Time Required (Hours/day)*					
		During Male Migration			*During Male with Female Migration*		
		Before Migration Work (hr/day)	*After Migration Work (hr/day)*	*Per cent Change*	*Before Migration Work (hr/day)*	*After Migration Work (hr/day)*	*Per cent Change*
Domestic work	7.00	5.50	7.50	26.67	7.00	4.50	–35.71
Child care	2.00	1.50	2.50	40.00	1.50	1.00	–33.33
Social participation	1.00	1.00	1.00	0.00	1.00	0.50	–50.00
Animal husbandry	1.00	0.50	1.00	50.00	0.50	0.00	–100.00
Agriculture or other work	6.00	5.00	6.00	16.67	5.00	8.00	+60.00
Other activities(sleeping, feeding, leisure time etc.)	7.00	8.50	7.00	–75.00	9.00	10.00	+11.11

During male migration the percentage change in time required for doing different works/activities like domestic work, child care, social participation, animal husbandry, agriculture/other work and other activities was +26.67, +40.00, +00.00, +50.00, +16.67 and -75.00 per cent respectively. It might be because of the fact that in the absence of male the responsibility of female towards family members and their participation in different activities is increased.

Similarly, during male with female migration the percentage change in time required for doing different works/activities like domestic work, child care, social participation, animal husbandry, agriculture/other work and other activities was -35.71, -33.33, -50.00,-100.00, +60.00 and +11.11 per cent respectively. It might be due to that migrated females are more engaged in their wage earning so that they devoted more time for the activities such as agriculture and give comparatively less time for family related and social activities.

Parganiha (2002) noted that workload of male during migration increased to the extent 37 to 200 per cent on activities such as agricultural and domestic work. But there was no effect on their social participation. While for female, their work load was reduced from 11 to 50 per cent on activities like domestic work, child care and time for rest. The workload on activities like agricultural work, social participation and other works was increased up to 25 to 50 per cent.

Conclusion

Migration is not harmful in terms of migrants because if migrants earn money without losing health, social reputation, poverty and education of children. Migration occurs because individuals search for money, food, employment and security outside their usual habitation. In Chhattisgarh state there is ample scope for employment through MGNREGA and assured food security through public distribution system (PDS). That restricts the scope of migration in the state. In Chhattisgarh only those people have migrated who are resource poor and don't have any employment opportunity in the off season or those people who are become habituated and have high level of aspiration. That's why here seasonal or temporary migration is prevalent.

In this study data were collected from command area. that's why here rate of migration is low (28.20 per cent).So, it may possible that the area which is not come under the command area the rate of migration is comparatively high then the command area.

This seasonal employment can be check to some extent if we increase the cropping intensity in the local vicinity of the migrants. This cropping intensity can be increased if assured irrigation facility is provided in the tail part of command area and rainfed area of the Chhattisgarh. So we can say that cropping intensity is negatively related with migration positively related with employment generation.

References

Anonymous. 2007-2008. National Sample Survey Office (NSSO).

Anonymous, 2013. International convention on the protection of the rights of all migrant workers and members of their families. Article 2.1. Adopted by General Assembly resolution 451158 of 18 December 1990. In force since 1 July 2003.

Anonymous, 2014a. Migrations country wise". Retrieved 7 June 2014.

Anonymous. 2014b. Migration in Chhattisgarh; Features, Trends and Policy Challenges with reference to the Inter-State Migrant Workmen (Regulation and Conditions of Service) Act, 1979.

Anonymous. 2015a. http://www.firstpost.com/business/nrega-and-after-large-scale-migration-of-rural-masses-is-back-in-chhattisgarh-2183681.html.

Anonymous. 2015b. http://timesofindia.indiatimes.com/india/Mass-exodus-100000-villagers-migrate-from-Chhatisgarh-in-3-years/articleshow/46818527.cms.

Anonymous, 2016. https://en.wikipedia.org/wiki/Human_migration.

Choubey, C. 2000. Study of factors responsible for low productivity of grasspea (*LATHYRUS SATIVUS* L.) under utera system in Chhattisgarh plains. M.Sc. (Ag.) Thesis, Raipur (C.G.).

Deshingkar and Akter. 2009. Migration and Human Development in India. United Nations Development Programme Human Development Reports Research Paper.

Jain, P.C. 1989. Emigration and Settlement of Indians Abroad', Sociological Bulletein, Vol.38, No.1.

*Lee, E.S. 1966. "A Theory of Migration" 3. University of Pennsylvania: 47–57.*JSTOR 2060063.

Parganiha, O. 2002. Impact of migration on livelihood system of the farmers in Raipur district of Chhattisgarh. M.Sc. (Ag.) Thesis, Raipur (C.G.).

Sharma, M.L., Hossain, M., Khan, M.A and Rathore, A.L. 2002. "The 2000 drought in Chhattisgarh, India: Impact on livelihoods and people's coping strategy". Paper presented in *'International Rice Research Congress* (IRRC-2002) *at Beijing* (*China*) from September 16-20, 2002.

Sharma, M. L, Rathore, A. L., Gupta, S. P. and Singh, A. K. 2009.Final report on baseline survey for first three batches of irrigation systems selected under CIDP.

UNFPA, 2013."International Migration 2013 (wall chart)".

United Nations, 2013a. World Migration in Figures: A joint contribution by UN-DESA and the OECD to the United Nations High-Level Dialogue on Migration and Development.

United Nations, 2013b. Trends in International Migrant Stock: The 2013 Revision.

Chapter 21

Empowerment of Hilly Women through Entrepreneurship in North-west (NW) Himalayas

N.A. Vijay Avinashilingam

Krishi Vigyan Kendra, ICAR-VPKAS,
Kafligair – 263 628 Bageshwar, Uttarakhand
Present Address: Div. – VI, ICAR-CAZRI,
Jodhpur – 342 003, Rajasthan

The past century evidenced a series of radical changes in agriculture reforms. Transformation in positive direction leads to development. The development in turn depends on effective utilization of available man power and natural resources. The objective of the human resource development emphasize that the men and women should enjoy long, healthy and creative lives with gender equity and equality. In an increasingly open global economy, Uttarakhand, a hilly state, which was carved out from the then U.P. of our country is bestowed with diversified agro-climatic features. This varied agro-climatic features nurture the hilly people to grow economically important medicinal and aromatic plants along with the cultivation of valuable minor millets. Keeping the basket of opportunities still the majority male population migrates to nearby cities in search of job opportunities leaving their women in the village to look after farming, animal husbandry and other household chores. In spite of being the majority in most of the district'softhis state the patriarchal nature of the society, gender-based division of labour, discrimination in resources allocation and remuneration have perpetuated the male dominance and chauvinism in Indian society given her the secondary status in hierarchy.

Keeping in view the above, developing women's entrepreneurship is essential for their empowerment. Even though the constitution guaranteed equality for women but still it is yet to be achieved in this hill state. Increasing independence of women socially, economically, psychologically and spiritually enhances their motivation, attitudes, values and knowledge. Inculcating entrepreneurial competencies (behavioural, entrepreneurial launching and enterprise management), entrepreneurial skills and enterprise building measures through training modules and intervention is the need of the hour to connect women from "Production to Consumption". This untapped intervention will definitely bring a paradigm shift in the status of hilly women in society from dependent consumer to producer and owners of enterprises.

Keywords: *Entrepreneurship, Competencies, Transformation and empowerment.*

Introduction

Uttarakhand, the 27th State of the Indian Republic was created from the Himalayan and adjoining districts of North Western U.P. It has a total area of 53,483 Km^2, of which more than 80 per cent is mountaineous and 65 per cent is covered by forest. Spread over twenty percent of the land area of the world, hills are inhabited by 10 per cent of the world's population and another 40 per cent live in the foot hills and its neighbourhood. Having wide variation in the climate and vegetation hills and mountains are considered to be rich source of bio-diversity and possess enormouos potential for sustaining livelihoods. Uttarakhand is an agrarian state and about 80 per cent of the population in this state depends on agriculture for its livelihood. Out of the total available land 12 per cent of the land is irrigated and about 64 per cent is fed by natural springs. The topography is characterised by sandy soils. Agriculture is being the mainstay, the improved scientific cultivation practices and technologies are still not reached to the majority of the farmers making the state to be highly economically backward. With these diverse agro-climatic situations, the agriculture scenario in hills is totally upside down when comparing to the plains, where they practice commercial agriculture. Even though the state's majority of people practice farming as their primary occupation the share it contributes to the country's total areaand production is very small. Despite about 75 – 85 per cent of the population dependent on the agriculture for their livelihood but the contribution of agriculture to state GDP is about 22.4 per cent. The production and productivity for different crops also vary greatly between plains and hills. The wheat and rice productivity in hills is 13.2 and 12.36 q./ha when compare to the plains which is 30.45 and 27.49 q./ha, respectively (Utttarakhand State Planning Commission, 2007). Further the seed replacement rate for hills is only 3 – 4 per cent whereas in plains it stood for 15 – 20 per cent.

Uneven excessive rainfalls coupled with hailstorms, prevalence of low temperatures throughout the year, high soil erosion, frequent avalanches and earthquakes force the farmers of Uttarakhand in search of the occupation which could provide supplementary income and sustain their family when their primary occupation fails. Due to the above problems, the male population of most of the families migrates to nearby cities/towns in search of job opportunities leaving their women in the village to look after farming, animal husbandry and other household chores. So in most parts of rural Uttarkhand it is observed that women contributes workforce up to 90 per cent in agriculture and animal care. Further the available data reflects that women form the core working force and their contribution rate is much higher in comparison to several states. In spite of being majority in most of the district's of this state the patriarchal nature of the society, gender-based division of labour, discrimination in resources allocation and remuneration have perpetuated the male dominance and chauvinism in Indian society given her the secondary status in hierarchy. Their involvement in farming, animal husbandry and being the saviors of protecting forest in this state can't be forgotten. They themselves proved as equal to their counterparts in many occasions. Further the dedication shown by them in upbringing their wards reflects their total devotion to the family. Therefore it could be rightly said most of the times in absence of their male they stood as a strong pillar in support of family members. Even though they have been actively engaged from farm to field, their status and standard were never raised. Despite being the back bone of the state's economy their potential is still untapped.

A relook is required to the so called invisible workers to visible workers, inside and outside the family. In the era of modernization and globalization the roles performed by female is gradually changing. During the past decades, women actively associate themselves in agriculture and allied activities, whereas during the recent years they started thinking how to actively contribute along with their counterparts in profession and livelihood income generation activities. Now the time had cane and they started looking various avenues and opportunities to empower themselves to improve their standard of living. Education plays an imminent role in empowering hilly women. In empowerment, education is considered to be the right eye then definitely employment must be the left eye. Both are almost inseparable. In employment working for others as well as giving work for others are totally mutually exclusive categories. The employee who works for others earn money as a salary would provide benefit for his family whereas if she/he provide employment (through entrepreneurship) for others indeed help in development of her/his family along with the other families. In this century the role played by women in economic development is inevitable. It has been often recognized that their integration in mainstream is crucial/critical factorin gender mainstreaming. Steps are being taken by various levels in promoting women entrepreneurship. Entrepreneurship is considered as one of the important driver contributing to the development of the society. There are certain identified qualities, traits and competencies required for a good entrepreneur. When the women folks are imparted with the above qualities along with entrepreneurial education and knowledge, they will definitely prove to be a potential force for the development of country in general and Uttarakhand in particular. Keeping in view the above, developing women's entrepreneurship is essential for their empowerment. Even though the constitution guaranteed equality for women but still it is yet to be achieved in this hill state. Increasing independence of women socially, economically, psychologically and spiritually enhances their motivation level and provide confidence in enterprise building.

Empowerment

The term empowerment roots from American Community Psychology and is associated with the social scientist Julian Rappaport (1981). Empowerment is a multidimensional process, which refers to measures designed to increase the degree of autonomy and self-determination in people and in communities in order to enable them to represent their interests in a responsible and self-determined way, acting on their own authority. (*Wiki Pedia*). It also involves individuals or group of individuals to realize their full identity and powers in all spheres of life. The Government always stressed the need for gender empowerment in Five Year Plans. Gender Empowerment is the empowerment of people of any gender, while conventionally being considered for the empowerment of women (*Wiki Pedia*). There are various aspects of women empowerment and those aspects of empowerment which made a visible impact among the women are briefly discussed.

Social Empowerment

It is a process of creating a favourable environment that provides easy access and minimum basic services to the young people to realize their full potential in their life situation. It also emphasis that the group approach is the more successful approach than the individual. It suggests taking steps in mobilizing poor women into Self Help Group's (SHG's) through the convergence of service, offering them a wide range of economic social options, along with necessary support services, to enhance their joining capabilities. (Premlatasingh, 2009)

Economic Empowerment

It is the capacity of women and men to participate in, contribute to and benefit from growth process in ways that recognize the value of their contributions, respect their dignity and make it possible to negotiate a fairer distribution of the benefits of growth. It increases women's access to economic resources and opportunities including jobs, financial services, property and othe productive assets, skills development and market information. It requires bold and sustained action to advance women's opportunities and rights and to ensure that women can participate and be heard. UNDP advances women's economic empowerment by:

- ✰ Advocating for increased recognition, reduction and redistribution of women's unpaid care work;
- ✰ Supporting research of customary law to help women gain access to land and participate in decision-making form
- ✰ Supporting women's entrepreneurship through training in production skills and techniques, business management and functional literacy;
- ✰ Helping to certify businesses that adhere to equality standards in the workplace;
- ✰ Supporting efforts to strengthen women's legal rights to property.

Psychological Empowerment

It is defined from the perspective, that the psychologically empowered state was considered to be a cognitive state characterized by a sense of perceived control, perceptions of competence, and internalization of the goals and objectives of the organization.Thus empowerment is thus considered as a multi-faceted construct reflecting the above three dimensions.

Spiritual Empowerment

Spiritual Empowerment and Self-Awareness hold the key to freedom and raising the status of women in society. Spirituality is all about awakening the 'consciousness' that we all humans have been gifted with. The consciousness to recognize the 'truth' about self, about relationship of self with the people and the nature around. This consciousness helps us to understand and become aware of ourselves and all things that have an impact on our well-being and our inner peace and happiness. This awareness is empowering because, it is the beginning of taking care of our own self as well as understanding others in a better way. Ideally this process must begin for the female from her younger years onwards to think for themselves and think about what they really want to do in future. Making available the examples and histories which could inculcate them the good habits and make them to rise above the normal human beings to take calculated risk. The teaching institutions and society in which they live must bring a positive atmosphere. Women, who comprise half of the population don't walk with us side by side will leads to the nation development half way. This decade the major thrust has been given to the entrepreneurship. So empowerment ofwomen through entrepreneurship is the need of the hour in today's context.

Empowerment through Entrepreneurship

Group formation caught enough attention during the IXth Five Year Plan (FYP). It continually got support during Xth FYP and the XIth FYP (2007-12) has recognized for the first time that women are agents of economic and social growth. The multi-faceted approach of the XIth plan, focused women before introducing any interventions and mentioned they must

(a) provide women with basic entitlements

(b) address the reality of globalization and its impact on women

(c) ensure an environment free from all forms of violence against women physical, economic, social and psychological

(d) ensure the participation and adequate representation of women at the highest policy levels

(e) strengthen existing institutional mechanisms and create new ones for gender mainstreaming and effective policy implementation.

Women entrepreneurship has been recognized as an important source of economic growth in our country. Even though considering the huge population of our country, there are fewer women entrepreneurs but the impact they create among their family and community is much wider. Empowerment strategies must start with enterprise building measures and inculcating entrepreneurial skills and competencies. Entrepreneurship in women could be developed through various training modules and interventions.

Table 21.1: State-wise Distribution of Small Scale Industrial Units Managed/Owned by Women Entrepreneurs

Sl.No.	*State/Union Territory*	*No.of Enterprises Managed by Women*	*No. of women SSI Units*	*Percentage to Total*
1.	Kerala	137561	139225	13.09
2.	Tamil Nadu	130289	129808	12.20
3.	Karnataka	101264	103169	9.70
4.	Maharashtra	80662	100670	9.46
5.	Andhra Pradesh	77347	77166	7.25
6.	Uttar Pradesh	54491	72667	6.83
7.	West Bengal	71847	69625	6.55
8.	Madhya Pradesh	62351	68823	6.47
9.	Gujarat	55361	53703	5.05
10.	Bihar	38170	49443	4.65
11.	Orissa	33274	38233	3.59
12.	Rajasthan	29785	36371	3.42
13.	Punjab	30190	29068	2.73
14.	Delhi	13368	14383	1.35
15.	Assam	111891	11757	1.11
16.	Manipur	9168	10745	1.01
17.	Chhattisgarh	11766	10034	0.94
18.	Haryana	10087	9620	0.90
19.	Uttaranchal	8706	8804	0.83
20.	Jharkhand	7271	7865	0.74
21.	Jammu and Kashmir	5640	5742	0.54
22.	Himachal Pradesh	3515	3722	0.35
23.	Mizoram	3076	3700	0.35
24.	Meghalaya	3658	3580	0.34
25.	Chandigarh	2059	2243	0.21
26.	Pondichery	1089	1065	0.10
27.	Tripura	631	863	0.08
28.	Goa	677	810	0.08
29.	Daman,Diu,Dadra and Nagarhaveli	167	213	0.02
30.	Nagaland	207	179	0.02
31.	Arunachal Pradesh	131	150	0.01
32.	Andaman and Nicobar Island	53	110	0.01
33.	Sikkim	30	98	0.01
34.	Lakshadweep	61	67	0.01
	All India	**1095843**	**1063721**	**100**

Source: SSI and Science Tech Entrepreneur, May 2007, Technopreneur.

Status and Statistics of Women Entreprenerus in India

Compared to the previous centuries the recent century has seen a remarkable change in development of entrepreneurship among women. However, with the passage of time, the British presence has the effect of generating self-awareness among women and they made their presence felt in the freedom struggle. Women are equally competent enough to run the business successfully in absence of their father. In overall scenario of the country, the four southern states and Maharashtra account for over fifty per cent of all women small scale industrial units in the country.The enterprises owned and managed by women in India are about 11 per cent of all enterprises. The total number of women enterprises in the total Small Scale Industries was estimated at 10, 63,721. The estimated number of enterprises actually managed by women was 10, 95,483. In the states of Kerala, Tamilnadu, Karnataka, Maharashtra etc. the share of women employment was significantly higher than other states. About 13 per cent of the women enterprises were in the registered SSI sector and the remaining 87 per cent were in the unregistered SSI sector.

Table 21.2: Classification of SSI Units in India on the Basis of Registration

Sl.No.	*Characteristic*	*Registered Units (In Lakhs)*	*Unregistered Units (In Lakhs)*	*Total*
1.	Men	12.37 (13.08)	82.20 (86.92)	94.57 (100)
2.	Women	1.38 (12.93)	9.26 (87.07)	10.64
	Total	**13.75 (13.07)**	**9.26 (86.93)**	**105.21 (100)**

Note: Figures in brackets represent percentages of total

Source: Science Tech Entrepreneur, May 2007, Technopreneur

Over a period of time the growth of women entrepreneurs in our country was phenomenal. It was estimated that presently in our country, out of the total entrepreneurs, 11 per cent is occupied by women entrepreneurs. This is expected to grow at the rate of 25 – 30 per cent of entrepreneurial force in India would be women by 2020.A women entrepreneur is defined as an enterprise owned and controlled by a women having a minimum financial interest of 51 per cent of the capital and giving at least 51 per cent of the employment generated in the enterprise to a woman.Women entrepreneurship is to be studied separately. It is essential to make them aware of the whole entrepreneurial process and the required entrepreneurial skills and competencies to run it successfully.

Entrepreneurial Process

The process is a sequence of steps required to achieve the objective/goals. It is also a course of action that involves all functions, activities and actions associated with identifying evaluating perceived opportunities and the bringing together of resources necessary for the successful formation of a new firm to pursue and seize the said opportunities (Bygrave, 1997:2; Cornwall and Naughton, 2003:62) Once set up, the process of entrepreneurship becomes effectively a cyclicalprogression of opportunity targeting and making strategic decisions regarding theallocation of scarce

1 **SELF-DISCOVERY:**
Learning what you enjoy doing; examining your strengths and weaknesses. Examining work experience and relating it to potential opportunities.

2 **IDENTIFYING OPPORTUNITIES:**
Looking for needs, wants, problems, and challenges that are not yet being met, or dealt with effectively.

3 **GENERATING AND EVALUATING IDEAS:**
Using creativity and past experience to devise new and innovative ways to solve a problem, or meet a need, and then narrowing the field to one best idea.

4 **PLANNING:**
Researching and identifying resources needed to turn the idea into a viable venture. Preparing the research in the form of a written business plan. Preparing marketing strategies.

5 **RAISING START-UP CAPITAL:**
Using the business plan to attract investors, venture capitalists partners. This stage can involve producing prototypes or test-marketing services.

6 **START-UP:**
Launching the venture, developing a customer base, and adjusting marketing and operational plans as required.

7 **GROWTH:**
Growing the business: developing and following strategic plans, adapting to new circumstances.

8 **HARVEST:**
Selling the business and harvesting the rewards. For many entrepreneurs, this also means moving on to new ventures and new challenges.

***Source*: http://www.mvp.cfee.org/en/aboutentwhatisentprocess.html accessed on 21.4.2016.**

resources in pursuit of value adding opportunities (Glancey, 1998:18; Kodithuwakhu and Rosa, 2002:443).Gruber (2002:193) identifies three distinct stages, namely the pre-founding stage(opportunity identification and evaluation); a founding stage (business plan, resourcegathering, incorporation and market entry); and an early development stage (building thecompany and market penetration).Entrepreneurs appear to work through a process that we can observe and describe. Often entrepreneurs discover this process through trial and error, but you can shorten your learning curve by benefiting from the experiences of others. The self-explanatory diagram given below could illustrate the logical sequence of activities.

Entrepreneurial Development Programmes (EDPs)

EDP is essential for the women entrepreneurs who are willing to take up entrepreneurship. This programme is meant to develop entrepreneurial abilities by continuously motivating the people. By this training it is understood that the entrepreneur could solve the initial problems which they will overcome while taking up the entrepreneurial task. In other words, it refers to inculcation, development, and polishing of entrepreneurial skills into a person needed to establish and successfully run his/her enterprise. Thus, the concept of entrepreneurship development programme involves equipping a person with the required skills and knowledge needed for starting and running the enterprise. It is designed for 6-8 weeks duration, aims at training the interested women entrepreneur of conceiving an idea about the prevailing socio-economic scenario, planto identify the business opportunities, initiating means to find out the various assistance offered by the agencies and finally launching an economic activity or an enterprise which is technically feasible and economically successful.

Entrepreneurial Competencies

Nowadays the age old proverb *i.e.* entrepreneurs are born not made had been disproven. By inculcating the required knowledge, skill, attitude and behavioural competencies to every individual, it is very much possible for everyone to become a successful entrepreneur. There is a list of general competencies, *viz.* behavioural, enterprise launching and enterprise managing competencies, which by acquiring one can able to excel in entrepreneurship. "Entrepreneurial competencies" refers to the key characteristics that should be possessed by successful entrepreneurs in order to perform entrepreneurial functions effectively.Entrepreneurial competencies are those skills necessary for an entrepreneur that helps them to,

- ☆ Venture into an enterprise
- ☆ Organize and manage an enterprise ably and competently
- ☆ Realize the goal for which the enterprise is established

Behavioural Competency

- ☆ Initiative
- ☆ Systematic planning
- ☆ Creativity and Innovation
- ☆ Risk taking and Risk Management
- ☆ Problem Solving
- ☆ Persistence
- ☆ Quality performance

- Information Management
- Persuasion and influencing abilities

Enterprise Launching Competency

The entrepreneur puts lot of effort and in making a decision to take an initiative. Once he is decided to take an initiative he has to take the initiative ahead of others. So, mental acumen, different thinking and thinking speed is also an important factor in the success of the enterprise because if the entrepreneur is slow in implementing the initiative, chances are there that someone else will move forward and implement it.This calls for alertness, agility and dynamism on the part of the entrepreneur.

Enterprise Managing Competency

To manage an enterprise is considered as so difficult in forming an enterprise. The entrepreneur must be expert in problem solving competency. This is nothing but,

- Acquiring the necessary mindset which will help to understand that the problems are part of the process and start working the problems in the context of the process to which the problems belong to,
- Basic understanding of the phenomenon that normally problems are always caused
- Clear understanding of the fact that every problem has a solution and one must seed for the solution.
- Decision making ability to choose the relevant, contextual and pragmatic solution.
- Explore the alternate strategies
- Generating new ideas, products, services, visions so that the problems of similar nature don't occur in future.

Apart from inculcating the above competencies, it is equally important that the training modules and interventions designed must suit the needs of women's groups.In conventional lectures a lecturer or trainer, lecturing before a group of trainees and the participants merely listen without any interaction. Whereas the entrepreneurship training modules/approaches engages participants to interact, encourage the trainees to discover themselves in the learning process. This experiential learning along with above said entrepreneurial skills and competencies creates the circumstances and conditions in which trainees participate and identifythemselves as one of the potential entrepreneur. Through these trainings, they explore and share their own experience and knowledge, critically conceptualize and reflect upon solutions from which they will derive learning contents and ways for practical implementation of an enterprise. Therefore the discussions will not only encourage growth, creativity and development of a women, but also it transforms them into a self-disciplined and an integrated personality in pursuit of entrepreneurship excellence. This chapter definitely supports the women not merely by knowledge acquisition but also making them to understand what entrepreneurship is.This untapped intervention will definitely bring a paradigm shift in the status of hilly women in society from dependent consumer to producer and owners of enterprises.

References

Bygrave, W.D. (1997). The Portable MBA in entrepreneurship. New York: John Wiley and Sons.

Cornwall, J.R. and Naughton, M.J. (2003). Who is the Good Entrepreneur? An Explorationwithin the Catholic Social Tradition. *Journal of business ethics*, 44:61-75.

Glancey, K., Greig, M. and Pettigrew, M. (1998). Entrepreneurial dynamics in small businessservice firms. *International Journal of Entrepreneurial Behaviour and Research*, 4(3):249-268.

Gupta, H.S. (2006) Hill Agriculture: Status of Research and Production Potential. In: Gupta, H.S., Srivastva, A.K. and Bhatt, J.C. (Eds.) Sustainable Production from agricultural watersheds in North-West Himalayas. Pub. Director, Vivekananda ParvatiyaKrishiAnusandhanSansthan, Almora pp: 1-21

Gruber, M. (2002). *Transformation as a challenge – new ventures on their way to viableentities.* Pages193-201. [Online] http://www.igw.unisg.ch/rencontres/band2002/C_03_Gruber.pdf - [Accessed 2007/09/10].

http://repository.up.ac.za/bitstream/handle/2263/24173/03chapter3.pdf?sequence=4accessed on 21.4.16

http://smallb.sidbi.in/per cent 20/fund-your-business per cent 20/additional-benefits-msmes per cent 20/women-entrepreneurship accessed on 20.4.16

http://www.angelfire.com/tn/moonlodge/reikiref.html accessed on 20.4.16

http://www.mvp.cfee.org/en/aboutentwhatisentprocess.htmlaccessed on 21.4.16

http://www.oecd.org/dac/gender-development/womenseconomic empowerment.htm accessed on 20.4.16

http://www.researchandmarkets.com/reports/1719107/women_empowerment_through_entrepreneurship accessed on 20.4.16

http://www.undp.org/content/undp/en/home/ourwork/povertyreduction/focus_areas/focus_gender_and_poverty/women_s_economicempowerment.html accessed on 20.4.16

https://en.wikipedia.org/wiki/Empowerment accessed on 19.4.2016

https://eshipnotes.wordpress.com/topics/class-xi/entrepreneurial-competencies/ accessed on 21.4.16

http://www.dsw.org/uploads/tx_aedswpublication/ENTREPRENUERSHIP_TRAINING_MANUAL.pdf accessed on 21.4.16

Kodithuwakku, S.S. and Rosa, P. (2002). The entrepreneurial process and economicsuccess in aconstrained environment. *Journal of Business Venturing*, 17(5):431-465.

Kushwaha, R.K. and Verma, Atish (2015). "Dynamic role of women entrepreneurs in Indian economic development" *Global Journal of Multidisciplinary Studies*Vol – IV, Issue – 10 Sep, 2015 pp: 62 – 69

Menon, Sanjay T (1999) Psychological Empowerment: Definition, measurement and validation *Canadian Journal of Behavioural Science* July 1999 accessed on 20.4.16

Premlata Singh, (2009) "Empowerment and the Self Help Group Movement" published in CAFT Training Manual on Entrepreneurship Development in Agriculture held between 5th – 25th Jan, 2009 at Div. of Agril. Extn., IARI, PUSA, New Delhi pp:109 – 115

RashmiSingh, (2009) "Women Entrepreneurship Development" published in CAFT Training Manual on Entrepreneurship Development in Agriculture held between 5th – 25th Jan, 2009 at Div. of Agril. Extn., IARI, PUSA, New Delhi pp: 93 - 108

Sinha, DK accessed on 2016 Entrepreneurship Development Programmes: Meaning, Need and Objectives of EDP http://www.yourarticlelibrary.com/entrepreneurship/entrepreneurship -development-programmes-meaning-need-and-objectives-of-edp/40707/ accessed on 26.4.16

Utttarakhand State Planning Commission, 2007 Statistical Reports of Uttarakhand

Vinesh (2014) "Role of Women Entrepreneurs in India" *Global Journal of Finance and Management* Vol. 6 (5) pp: 473 - 480.

— *Recommendations* —

ISEE National Seminar on Extension Innovations and Methodologies for Market-Led Agricultural Growth and Development

Venue: RVSKVV, Gwalior (M.P.)
Date: February 26-28, 2015

A National seminar" was organized by Indian Society of Extension Education, IARI, New Delhi in collaboration with the Department of Extension Education, Rajmata Vijayaraje Scindia Krishi Vishwa Vidyalaya, Gwalior at the main campus Gwalior (M.P.) from February 26 – 28, 2015. About 300 participants consisting of eminent Agricultural Extension Scientists, Development Officers, Public and Private Government Officers, KVK Scientists, Policy makers, Administrators and Progressive farmers participated and taken part in deliberations on different thematic areas as follows:

1. Linkages and institutional innovations for market led agriculture
2. Extension innovations, methodologies and experiences in technology transfer, quality up-gradation, value addition and processing and marketing.
3. Rural entrepreneurship for market driven agriculture and livelihood systems.
4. Policy initiation for market-led- rural economy.
5. Information and communication technology for Agri-business management including marketing.
6. Gender main streaming – extension imperatives for market led agriculture.

7. Harnessing of indigenous technical knowledge and farmers' innovations for agro-marketing systems.

The Indian Society of Extension Education has been playing pivotal role in agriculture development of the country since its inception in 1965. It organizes National/International Seminars annually besides brainstorming on agricultural issues having national importance. Through these activities, the society has made significant contribution in agricultural development as well as initiation of many technology transfer programmes in the country. The ISEE seminar - 2014 was a step-forward effort to assess and analyze the progress of work in extension research, training and field extension and bring out recommendations for market led agricultural growth and suggestions for the improvement upon present and future extension programmes and activities. The outcomes of the seminar would lead to transform the Indian agriculture and make it highly viable and remunerative including agri-preneurship development among farmers, farmwomen, rural youth and convergence of public and private efforts for marketing and export.

Three days rigorous deliberations and discussions on the research findings and field based experiences concerning different thematic areas of the seminar have resultant many fruitful recommendations and suggestions have emerged from the Seminar for agricultural progress and prosperity. This would help in higher agricultural production and productivity, micro enterprise development for income and employment generation, strengthening research and extension systems and their linkages, higher post-harvest productivity and value added products, gender empowerment and capacity building towards entrepreneurships and household food and livelihood security.

The seminar was inaugurated by distinguished scientist and academician Dr. S. N. Puri, former Vice-Chancellor, Central Agricultural University, Imphal, Manipur on 26th February, 2015 at 11.30 AM in the Auditorium, College of Agriculture, Gwalior (M.P.). Dr. Baldeo Singh, President (ISEE), welcomed the Chief Guest and dignitaries and briefed about the National Seminar. Professor A.K. Singh, Vice- Chancellor RVSKVV, Gwalior and Petron of National Seminar presided over the function. The Chief Guest released the National Seminar Souvenir cum Abstract, the Indian Journal of Extension Education issue January to July, 2015 and an Extension book. The Life Time achievement award of the ISEE was given to the renowned Extension Scientist Prof. R.P. Singh by the Chief Guest. Dr. M.M. Patel, Head, Department of Extension Education and Local Organizing Secretary, National Seminar extended vote of thanks.

The Distinguished Scientist Dr Bangali Babu, Former National Director, NAIP, ICAR was the chief guest of the Valedictory function. Dr. Baldeo Singh, President (ISEE), welcomed the Chief Guest and dignitaries. Prof. V. Veerbhadraiah, Ex- Emeritus Scientist (UAS, Bangalore) and Ex-Vice President of the ISEE, presented the salient recommendations and suggestions of the seminar. Professor A.K. Singh, Vice-Chancellor, RVSKVV, Gwalior and Petron of National Seminar presided over the function. The Chief Guest distributed ISEE awards to the eminent extension scientists and extension professionals working in agricultural universities and ICAR institutes including Krishi Vigyan Kendras for their significant contributions in agricultural extension fields. Dr. Kartar Singh, Vice-President of Indian Society of Extension Education extended vote of thanks to all directly and indirectly involved and supported in successful organization of national seminar.

During the seminar, total seven thematic areas were deliberated under one Joint session and six technical sessions. Out of seven thematic areas of the Seminar, 32 recommendations were emerged from three days deliberations. The session wise recommendations / suggestions are given hereunder;

Major Recommendations

Joint Technical Session

Chairman: Dr. Baldeo Singh, Former Joint Director Extension (IARI) and

Vice Chairman: Dr. Eshwarappa, Director, RKVY (Karnataka)

A total of 60 papers including two lead papers were presented.

Lead Papers Recommendations

- ☆ FPOs (Farmer Producers' Organizations) to be promoted largely on all commodities covering all geographical area. Adoption of market led extension, methodology, precision farming technology, ICM package adoption by farmers, linkage between production and marketing etc.are critical for sustainability of FPOs.
- ☆ Linking FPOs with IT based services is essential.
- ☆ Development of village based integrated models for market led agriculture in different parts of India in collaboration with SAU/ICAR institutes including important NGOs is suggested. Imperatives for sustainability such as employment generation, food and nutritional security, gender equity, poverty alleviation, environmental sustainability institutional and political support are needed.

Farmers' Interaction

- ☆ Sh. Chhatrapal Pateria from Datia district emphasized the use of ITK in agriculture. He appreciated the efforts of KVK on NICRA initiative.
- ☆ Sh. Abhay Kher from ashok nagar district emphasized the organic farming in present era. He discussed about various initiative and awards received by him on different forums for organic farming and value addition.
- ☆ Sh. Shyam Lal from Shivpuri district narrated his achievements by adopting technologies disseminated by KVK. He also emphasized the role played by radio and mobile messaging in rapid transfer of technology.
- ☆ Sh. Ramesh Chand Rajawat from district Gwalior emphasised the role of media for transfer of technology. He emphasized on proper timing of telecast on the benefit of farmers. He is also producing and seeking organic manures.

Technical Session – I

Theme No. 1: Linkages and institutional Innovation for market led Agriculture

Chairman: Dr. G. Eswarappa

Co-chairman: Dr. Rajender Kaur

Rapporteurs: Dr. H. K. De and Dr. M. Shivamurthy

Out of 20 papers, 10 were presented in the session. Lead paper on "Institutional Innovations for marketing of farm produces – KVK experiences" was presented by Dr. H.K. De. He spoke on a few successful initiatives by KVK's for bringing buyer and seller together. With KVK's intervention successful market linkages could be established. He stressed on upscaling of such innovations.

- Dr. Shobhana Gupta emphasized KVK–ATMA convergence. She presented the impact of convergence at RSKVV in terms of increase in yield, delivery of advisories through portal, skill training to farmers etc.
- Dr. M Shirvamurthy discussed about strategies for making small farms productive as well as profitable. Appropriate technology for small farms and their marketability needs to be looked into.
- Dr. A. Prema talked about private public partnership (PPP) in rice in Kerala. IT enabled value chain management involving supply, mills, farmers have enabled farmers realizing higher price. She stressed on integrated procurement, processing and distribution of rice.
- Dr. D.D. Sharma presented a comparative picture of production led extension and market led extension. He stressed that farmers are required to make farm related discussion, keeping an eye on the market.
- Jolly Masih presented a study on supply chain management of fruit and vegetables in Delhi and NCR.
- Dr. N. Loganandan presented two case studies on direct marketing by farmers. Commodity value addition, doorsteps marketing with the guidance of KVK have yielded better returns.

Recommendations

- Some KVKs have initiated innovative measures for linking produces to the market. Such initiatives need to be up scaled.
- Farmers need to be educated about what to produce, how much, when and where, keep it in view the market.
- Value addition and door step marketing directly to consumers could also be considered in selected cases.
- Appropriate technology for small farms need to be developed. Small farms could be brought together as producer organization. This would minimize middle men and would help realize better price for their produce.
- KVK-ATMA linkage needs to be strengthened to the advantage of farmers.
- Private public partnership in value chain management of important crops is required. Integrated procurement, processing and distribution as in Kerala may be replicated.

Technical Session – II

Theme No. 2: Extension Innovations Methodologies Experiences in Technology Transfer, Quality Up-Gradation, Value Addition and Processing and Marketing

Chairman: Dr. K.N. Krishnakumar, Professor, Indian Institute of Forest Management, Bhopal

Co-chairman: Dr. Dandu Jagannadha Raju, Former Director EEI, Hyderabad and Former Registrar ANGRAU, Hyderabad

Rapporteurs: Dr. Dipak Kumar Bose, Associate Prof., Dept. of Agril. Extension, SHIATS, Naini, Allahabad
Dr. Biswarup Saha, Assistant Professor (S.S.), College of Fisheries, Central Agricultural University, Tripura

Under this theme fifteen papers were presented by the extension researchers from different part of the country. The papers were concerned with wide range of Extension Innovations and Methodologies in Market led Extension related to agriculture and allied sectors including fisheries, horticulture, watershed management, lac culture etc. The Scientists also shared their experiences with regard to the dissemination of these innovations to the farming communities and their responses and perceptions while adopting those innovations.Most of the papers emphasized participatory approaches in extension methodologies in technology generation, testing, dissemination and marketing of agricultural products.

Recommendations

The present day extension approaches needs to be group approach mode where good success cases reveals commodity based association/ farmers interest groups encourages the farmers to join together in receiving the appropriate technologies as well as in adopting those technologies collectively and in processing their produce, value addition and marketing for obtaining better returns.

- ☆ The extension personnel should get in touch with the day to day developments going on in the areas where they are working regarding advising the farmers in what to be grown? When to be grown? How to be grown? How much to produce? Where to sell? When to sell? How to sell? In order to get higher profit.
- ☆ The important extension methodologies in market led extension suggest that the capacity building with proper skill oriented approach will yield good results by inculcating the timely technical inputs to the farmers and the extension officers of agriculture and other allied sectors through KVKs, SAUs, and EEIs.
- ☆ ICTs may take a lead role in market led extension approaches as extension education has become high tech mode in reaching more number of farmers with location specific technologies using KVKs as base centres.
- ☆ Value addition approach by organising special exhibitions with value added products of various commodities emerging in those areas.

- ☆ More extension innovations are needed in the area of value chain systems, supply chain and marketing channel for location specific commodities.
- ☆ Proper planning and functional linkages of Research-Extension-Farmers-Market intelligence are required for securing the livelihood of the farmers.
- ☆ In fisheries sectors technician based approach may be introduced in our extension system as this approach is proved successful in the fisheries sector of West Godawari district of Andhra Pradesh where in the extension approach was totally being followed by the private fisheries companies by engaging simple technicians in getting and analysing soil and water samples then and there and advising farmers in the field condition itself as well as in supplying the needed pesticides and fertilizers bio-agents etc., and time to time updating the farmers with proper fisheries technologies and market intelligence based analysis at a farmer or group of farmers so as to motivate the farmers and assuring them in purchasing back the total produce of the farmers.

Technical Session - III

Theme No. 4: Policy initiatives for Market led Rural Economy

Theme No. 7: Harnessing of Indigenous Technical Knowledge and Farmers' Innovations for Agro-marketing systems.

Chairperson: Prof. V. Veerbhadraiah

Co-chairperson: Dr. M.L. Sharma

Rapporteurs: Dr. Surya Rathore
Dr. Lalit Kumar Tyagi

- ☆ Dr. Surya Rathore presented a lead paper on 'Boosting Rural Economics through Local Economic Development (LED): Policy initiatives. In order to boost rural economy, local economic development is required in the rural areas. This can be achieved by means of eco-tourism, agri-tourism, self-help groups, agri-preneurs, promotion of rural craftsmanship, organic farming and para-vets, etc. She also highlighted policy issues required in these aspects including various awards for promotion of eco-tourism, teaching traditional crafts in schools as vocational education, development of parternships between agri-tourism and value added business, etc.
- ☆ Dr. L.K. Tyagi presented a lead paper on 'A Critical Review of the Extension Science as Practiced in India with an Agenda for Reform'. In the background of the evolution and growth of the extension science as a discipline, he discussed some of the issues and challenges extension science is facing in the current professional environment in the NARS. Four kinds of issues were discussed: 1. administrative and image-related issues, 2. issues related to the teaching and curricula, 3. issues related to research and 4. Issues related to professionalism in Extension as a science in India. An agenda for reforming the discipline by much needed actions on various fronts was also outlined.

- ☆ Dr. M.L. Sharma presented his paper on 'Farmers' perspective assessment of migration pattern in Chhattisgarh: Policy issues for sustainable livelihood'. He gave overview of migration patterns among members of Water users associations of 12 districts of Chhattisgarh. He highlighted that by increasing the cropping intensity, the migration of rural folk to urban areas can be reduced.
- ☆ Dr. Neeraj Singh presented his paper on 'Challenges to extension in changing scenario of vegetable production'. He highlighted that human capital building in vegetable production is important. He highlighted several challenges to extension system in vegetable production. He also opined that use of ICT is required by extension personnel.
- ☆ Dr. M.C. Ahire presented his paper on 'Indigenous practices followed by the farmers for storage of food grains'. He gave an overview of various indigenous practices reported by various researchers in food grains storage. He opined that many of these practices should be promoted among the farmers.
- ☆ Dr. R.K. Singh presented his paper on 'Method of seed treatment and seed sowing practices in ancient India and their relevance in present context'. He elaborated various seed sowing practices from ancient literature.
- ☆ Chairman remarked that most of the papers were less relevant for the themes of the session except the paper on Rural Economics through Local Economic Development. He remarked that the paper on 'A Critical Review of the Extension Science as Practiced in India with an Agenda for Reform' should have been put in the plenary session as this has very relevant for all the participants.

Technical Session – IV

Theme No. 3: Rural Entrepreneurship for Market Driven Agriculture and Livelihood System.

Chairman: Dr. V. Veerbhadraiah

Co-chairman: Dr. V.B. Dixit

Repportures: Dr. K.K. Jha
Dr. Bineeta Satpathi

In total, there were nine presentations in the session.

1. Income and Employment Generation Through Secondary Agriculture: A Case of Successful Rural Woman of Faridabad District of Haryana by Dr. V. P. S. Yadav

- ☆ Promotion of SHG's fostering economic activities like value addition in Soybean with the help of KVK interventions may be taken up.
- ☆ Focus to be made on secondary agriculture for market led extension.

2. **Producer collectives for value chain intervention and linkage with Market by Dr. R. N. Padaria**
   - ☆ Focused on Dhanya Mahotsav model for marketing and farmer to farmer knowledge sharing.
   - ☆ Direct marketing linkage was stressed upon with logistic improvement.
   - ☆ Linking farmers with collective action.
   - ☆ PPP based value chain development in linseed marketing.
   - ☆ Emphasis on socio-metric analysis and community based extension services.

3. **Broiler production for developing rural entrepreneurship in and around Jammu city of J&K by Dr. S. Dwivedi**
   - ☆ Broiler production is an effective instrument for rural entrepreneurship.

4. **Entrepreneurial behavior of pineapple growers in Nagaland - An analysis by Dr. K. K. Jha**
   - ☆ Public-private community partnership model may be promoted for market led extension in case of pineapple crop in Nagaland.

5. **Market orientation through entrepreneurial behavior in secondary agriculture by Dr. B. Satpathy**
   - ☆ Promotion of value addition in groundnut for better market opportunities.
   - ☆ Strengthening farmers club at village level for rural marketing with examples from the district.

6. **Community based PPP model for women agri-preneurship development by Binoo P. Bonny**
   - ☆ PPP community model for marketing of agri-produce may be promoted.

7. **Perception analysis of agricultural experts on setting priority for development of agri-business in vegetable sector by Dr. N. Singh**
   - ☆ Web based module development for integration of vegetable marketing.

8. **Beekeeping – Demand driven market led extension by Dr. Esakamuthuu**
   - ☆ PPP module may be emphasized for encouraging beekeeping as a potential enterprise.

9. **Strategies to foster women entrepreneurship through SHG by Dr. Seema Tyagi**
   - ☆ SHG as a powerful tool for women empowerment emphasizing on literacy and organizing members into leaders.
   - ☆ SHGs can also constitute to market led extension.

Recommendations

- ✰ Public-private community partnership model may be promoted for market led extension.
- ✰ Value addition of agricultural products with interventions of KVKs for marketing may be promoted.
- ✰ Branding of agricultural produce can be an asset for marketing as in case of Dhanya Mahotsav Model.
- ✰ Public-Private partnership value chain development and emphasis on socio-metric analysis with community based extension may be promoted for market led extension.
- ✰ Promotion of agri-preneurship with strengthening of SHG may be a powerful tool for market led extension emphasizing on literacy and organizational management of groups.
- ✰ Farmers' club may be promoted with support from central, state and the local communities.

Technical Session – V

Theme No. 5: ICT for Agri-business management including marketing.

Chairman: Dr. M. S. Nataraju

Co-Chairman: Dr. Milind C. Ahire

Rapporteurs: Dr. C. Karthikeyan
Dr. Rupasi Tiwari

A total of eight papers were presented in the session.

(**Speakers:** Dr. K. N. Krishna Kumar, Dr. C. Dr. C. Karthikeyan, Dr. Rupasi Tiwari, Dr. Y. D. Mishra, Dr. Sarvesh, Dr. Kumari Manisha, Dr. V. K. Yadav and Dr. Nagashree.)

Recommendations

- ✰ Modern and information rich extension system requires greater integration of different key stakeholders and sue of ICT platforms to promote market led extension in India.
- ✰ Integrating the services offered by Kisan SMS portal of the GOI and the e-velanmai (*i.e.*, e-agriculture approach of extension performed by the Block Technology Managers (BTM), Asst. Technology Managers (ATM) of ATMA has been proved to render cost effective and quality extension service in Tamil- Nadu. Such attempt can be replicated in other states too.
- ✰ The potential of ICT tools *viz.*, touch screen kiosks, tablets, portals, expert systems, educational videos films, decision support software must be harnessed for their effective applications in agricultural extension services for the benefit of farmers.
- ✰ When preparing the contents for various media care must be taken to give emplacing to location specific technologies, local language, culture and needs of the targeted audience.

☆ An integrated agro advisory and trading platform exclusively for agriculture must be created and properly regulated by public agency for the benefit of all stakeholders in agricultural sectors.

Technical Session – VI

Theme No. 6: Gender Mainstreaming-Extension imperatives for Market-led Agriculture

Chairman: Dr. U. S. Gautam, ZPD – Zone IV (ICAR), Kanpur

Co-chairman: Dr. J. P. Yadav

Rapporteurs: Dr. Binita Rani
Dr. M. Shivamurthy

Lead Papers

1. Emerging role of KVKs for woman improvement presented by Dr. U.S. Gautam

Dr. Gautam proved, through this module, improvement in the socio-economic status of the farm woman. He advised that Home Scientists should choose an intervention and activities which are appropriate for solving the problems of farm woman and improving their living standards and sources of revenues. The study recommended that the Home Scientists should choose an intervention and activities which are appropriate for solving the problems of farm woman on drudgery reduction, enhancing efficiency, income generation, nutritional security, value addition and processing for improving their Livelihood.

2. Identifications and analysis of intervention programmes for gender mainstreaming in agriculture presented by Dr. Rameshwari Pandya

She recommended female children education as campaign or inviting participation of media, NGOs etc. she also emphasized to encourage financial institutions for providing small loans.

Other Papers

1. A study on various aspects of gender inequity in India by Aparna Jaiswal

There is need to ensure women equal access to basic education and skill development, health care, child care etc. and also to prevent crime against them and providing safe environment to them.

2. Importance of gender mainstreaming in agricultural marketing by Lata Sati

She concluded about improving the status of woman socially and economically by providing them powers of discussion making, support services like gender sensitively trainings, information and awareness about refutes and duties of woman etc.

3. Impact of agricultural labour shortage on woman in rural areas by Sarda Prasad

He discussed about female work participation and labour shortage in agriculture and suggested for qualitative improvement among them through health care, equalization in wage rate and discussion powers.

4. Role of community Radio in empowering farm woman through mushroom production by Binita Rani

There is need to organize entrepreneurship development training programmes for empowerment of farm women.

5. Empowerment of hilly women through entrepreneurship in North-west (NW) Himalayas by Vijay Avinashilingam

He emphasized for entrepreneurship in which he recommended empowerment of women through intensive capacity building and leadership development trainings.

Recommendations

- ☆ The Home Scientists should choose on the problem base an intervention and activities which are appropriate for on drudgery reduction, enhancing efficiency, income generation, nutritional security, value addition and processing for improving their Livelihood.
- ☆ Female children education as campaign or inviting participation of media, NGOs etc. she also emphasize to encourage financial institutions for providing small loans.
- ☆ There is need to ensure women equal access to basic education and skill development, health care, child care etc.
- ☆ He emphasized for entrepreneurship in which he recommended empowerment of woman through intensive capacity building and leadership development trainings

Final Recommendations of the National Seminar

1. Development of village based integrated models for market led agriculture in every KVK district in different parts of India in collaboration with respective SAUs/ICAR institutes. Imperatives for sustainability such as employment generation, food and nutritional security, garden equity, poverty alleviation, environmental sustainability, institutional and political support are needed.
2. Farmers from Datia district emphasized the use of ITK in agriculture. He appreciated the efforts of KVK on NICRA initiative.
3. Appropriate technology for small farms need to be developed. Small farms could be brought together as producer organization. This would minimize middle men and would help realize better price for their produce.
4. Proper planning and functional linkages of Research-Extension-Farmers-Market intelligence are required for securing the livelihood of the farmers.

Public-Private- community partnership model may be promoted for market led extension.

5. Value addition of agricultural products with interventions of KVKs for marketing may be promoted.
6. Public-Private partnership value chain development and emphasis on socio-metric analysis with community based extension may be promoted for market led extension.
7. The Home Scientists should choose the problem based on an intervention and activities which are appropriate for drudgery reduction, enhancing efficiency, income generation, nutritional security, value addition and processing for improving their Livelihood.
8. The emphasis was given for development of women entrepreneurships. The empowerment of woman through intensive capacity building and leadership development trainings was recommended.